"We extract solid and liquid min[eral] ... earth to produce materials useful t[o] ... production is often inimical to ou[r] ... on this planet. David Howe's fine account, *Extraction to Extinction*, explains much – a lyrical and questing narrative of how humans have used and abused natural resources down the ages – both to their great advantage and, now we realise, to their potential detriment. The author brings a double-edged pen to bear – long-brewed technical knowledge combined with an easy story-teller's acumen, fluency and wisdom. Just about any reader of his account will come away better informed and a wiser person to engage with the major conundrums of this material age."
MICHAEL LEEDER, PROFESSOR EMERITUS, UNIVERSITY OF EAST ANGLIA, author of *Measures for Measure: Geology and the Industrial Revolution* (Dunedin)

Praise for David Howe's previous books:

"Written in a tremendously readable narrative style (think *Mountains of the Mind* by Robert Macfarlane) ... skilful ... compelling." *The Great Outdoors*

"Chapters on mountains, ice and lakes, on farming and national parks ... set against wider cultural importance ... very accessible ... [yet] holds a considerable weight of authority." *Country Life*

"An ambitious and enjoyable biography of a landscape." *Cumbria Life* BOOK OF THE MONTH

"Interesting ... links the landscape with the scientists ... forensic details of the geology and geomorphology of the Lakes ... detailed biographies on the lives and loves of the Romantic poets." *BBC Countryfile Magazine*

EXTRACTION
TO EXTINCTION

RETHINKING OUR RELATIONSHIP
WITH EARTH'S NATURAL RESOURCES

DAVID HOWE

Published by Saraband
Digital World Centre, 1 Lowry Plaza,
The Quays, Salford, M50 3UB
www.saraband.net

ISBN: 9781913393274

1 2 3 4 5 6 7 8 9 10

Printed and bound in Great Britain by Clays Ltd, Elcograf S.p.A.

Author's note: This book deals in many different scales and substances. For the reader's ease, I have chosen to use what I believe to be the most readily understood unit of measurement in each case, rather than to clutter the book with alternatives or render everything in one system of units that will then need to be translated by the reader into something meaningful to them.

In memory of
Roy Coleman
and
Henry Emeleus

Contents

Chapter 1

Rocks and Resources

I was standing on Alderley Edge when I first wondered about it. We'd been exploring the New Red Sandstone rocks that rise abruptly above the Cheshire Plain. It was our first field trip with Mr Coleman. Roy Coleman, with his deadpan humour, was a charismatic teacher. The eight of us were sixth-form students in his geology class. He was helping us to get a sense of what the world was like around these parts 250 million years ago, a world of desert sands and river flood plains straddling the Tropic of Cancer, far to the south of where we were standing today.

The highlight of the morning was when we found a few tiny crystals of malachite. Malachite is a copper carbonate mineral, bright green in colour, vibrant and beautiful. It was one of the minerals that Bronze Age people discovered and used to smelt copper, before mixing it with tin to make bronze around 5,000 years ago. And here were malachite crystals embedded in a sandy rock outcrop a few hundred feet above the flatlands that stretched to the distant outlines of hazy Manchester, a dozen miles to the north. Millions of years after their burial and compaction, super-hot, mineral-rich waters had percolated through these lithified desert sandstones. They flowed along fault lines, fissures and cracks. As the heated waters slowly cooled, searching fingers of malachite veins began to crystallise.

The veins had enough copper in them to encourage local Bronze Age people to mine the metal. Indeed, there was so much copper ore in these rocks that mining continued on and off right up until the early twentieth century. All that remains today, though, are the labyrinthine old mine shafts that partly inspired Alan Garner to write his wonderful fantasy novel for children, *The Weirdstone of Brisingamen*, a story about wizards and witches.

The thing I was pondering as I looked at the malachite crystals was a different kind of magic and wizardry. What a trick, I thought, to turn a lump of rock, albeit stained green and blue, into a shiny red metal that could be fashioned into arrows and axes. It was all very well feeling excited about finding vivid green crystals in a rusty red rock, but I couldn't imagine it ever crossing my mind that somehow it could be turned into something else, something so very different – a brightly coloured metal that could be shape-shifted and beaten into ornaments and shields, pins and tools. How did they do it? How did they *know* how to do it?

At the end of the trip, we piled into the school bus and drove down the steep hill. We passed large houses hidden by high walls, tall trees and heavy gates that thirty-odd years later would become the homes of Manchester's super-rich footballers, who may, or may not, have been aware of Alderley Edge's history of desert sands, copper mines, wizards and witches. We drove through the town itself in a vehicle made of steel powered by petrol, along a tarmac road beside slabs of concrete paving, past shops and plate-glass windows, by houses built of stone and brick, and on, along the outskirts of Manchester's 'Ringway' Airport, just as an aluminium-framed Vickers Viscount turboprop passenger plane was coming in to land.

Iron, steel, aluminium, glass, concrete, stone, brick and almost every bit of material stuff I could see, both inside the bus and outside, had started life as a rock of one kind or another. What a feat, I thought. Rocks under the ground making so much of the stuff we see around us. How extraordinary. How clever.

And there I left the thought for the next fifty-five years, unremembered, until recently. I was on my computer, the one on which I'm typing this now, seeing which was the shortest or quickest route from Norwich to the south of Manchester. The one I eventually favoured took me across country, over the Pennines, through Buxton and Macclesfield, by Alderley Edge, on to Wilmslow, along the perimeter of Manchester International Airport, before reaching Altrincham.

Alderley Edge! Mr Coleman. Triassic New Red Sandstones. And malachite; do I still have that little chunk of rock with the malachite crystals? Then, in that strange way in which the mind works, I remembered my half-century-year-old thought. For some reason it burst into the present with unexpected force. I began to feel quite excited.

Yes! Good heavens. Everything in this room, too, I thought, started life in the earth, as a rock or a mineral vein, a layer of an ancient seabed, or perhaps the remains of a 400 million-year-old volcano. The plastic casing around my computer screen, the steel radiator beneath the aluminium-framed windows with their panes of glass, the copper wiring carrying the electricity to my laptop, the rare earth metals that enable my mobile phone to perform its many wonders, the clays in the paint that brighten up the walls of this room. Before they all became this material stuff, they were the end product of some geological process, some rock-forming activity, some rock-weathering breakdown, some human intervention.

And then I calmed down. Humanity's ability to fashion nature to its own ends goes back tens of thousands of years. The first rocks that lent themselves to being shaped into useful objects and tools were ones that had a very fine grain, or in the case of the more glassy rocks such as black obsidian and pale grey flint, no grain at all. This meant that when you hit them and knapped them, they would fracture into 'conchoidal' pieces with razor-sharp edges. It was these sharp-edged flakes and chipped pieces that were prized. They could be used as arrowheads, axes, tools to scrape animal skins, and knives to cut meat. What was *in* the earth was becoming as important to our ancestors as what was growing *on* and wandering *over* the Earth.

Jump forward thousands of years to the present day, and our species' ability to transform nature so radically, so fundamentally, into the material world all around us is staggering, even magical, and not a little frightening.

Take the mineral silica, or quartz. It's made of silicon dioxide. The flints that were knapped in the Stone Age are mainly made of cryptocrystalline silica, which is to say, a rock whose crystals are so small you need a powerful microscope to see them. Many centuries later, people understood how to melt silica sands and make glass. Twentieth-century scientists took the magic even further. They learned how to grow tiny slivers of pure silicon. These are the microchips upon which most of our computers depend. Flint arrows, glass and silicon microchips, all based on the Earth's gift of tiny, sandy grains of silicon dioxide.

It's the 1980s. I'm in North Yorkshire overlooking a rugged landscape of rusty-brown rock. I pick up one of the broken chunks that litter the strewn slopes. It's heavy to the feel, heavier than I was expecting. By a series of processes developed over the centuries, we have somehow learned to turn this type of rock into molten iron and shiny steel. Further effort and determination found ways to fashion steel into girders and knives, car bodies and kitchen sinks. In the nineteenth century, mining these Cleveland Ironstones helped build the Teesside iron and steel industry, which played its part in forging the Industrial Revolution.

Travel seventy miles east. Take another fragment of rock. Its colour is a light, shiny grey. It shimmers with silvery-sheened crystals. Again, it feels surprisingly heavy. The hills of the Northern Pennines are riddled with old mines, many dating back to the nineteenth century and before. More ingenuity and, hey presto, we have turned the rock and its crystals of smooth-sided galena into bright silvery blobs of molten lead, ready to be cast into pipes, gutters and roof linings.

Nowadays, we mine, quarry, pump, cut, blast and crush billions and billions of tons of the rock that lies beneath our feet every year. We are digging into and scraping away the surface of our planet at an unprecedented rate. We are shifting so much of the Earth's crust, and to such an extent, that we are now far outstripping nature's annual capacity to move mountains and fill oceans through wear, tear and weathering. And when all of this rock has been crushed and carried to the furnaces and factories, the cunning of men and women reworks these earthly treasures into new, marvellous things. We transmute nature. We weave our rocky resources into the very fabric of the modern world in which we live, work and play.

* * *

As a species, Homo sapiens have been around for about 200,000 years. We had our origins in north-east Africa. Our hunter-gatherer lifestyle meant that for a long time we lived in harmony with nature. Wood was collected as a fuel for fires, which kept us safe and warm and on which we learned to cook our food. The wood from the trees quickly regrew. As well as burning it, men and women learned to carve and shape wood to make shelters, spears and tools for digging and grubbing. These demands on the forest were sustainable. New trees would soon grow.

For a long time, we also had a passive relationship with rocks. We could shelter inside them wherever there was a cave. We could pick up a pebble and weight a fishing line. Small lumps of rock could be used to crush the shells of nuts and seashore molluscs. All of this meant that the landscape would look no different before and after men, women and their families had passed through.

Although human beings are less passive nowadays, we still use many rocks and stones in their natural, unchanged state. The history of building with stone goes back thousands of years. The Egyptians began building their pyramids more than 4,600 years ago. The earliest are found at Saqqara, north-west of Memphis, which lies about 12 miles to the south of modern-day Cairo.

The most famous, iconic pyramids are found at Giza, the tallest of which, the Great Pyramid, rises 481 feet high. The pyramid complex lies on the south-western outskirts of Cairo. Most were built as tombs for dead pharaohs. The construction of these huge mausoleums still beggars belief. The outside stones of the Great Pyramid are blocks of locally quarried limestone, each weighing several tons. However, some of the internal stonework that forms the burial chambers is made of granite. These 25-ton or more blocks were quarried at Aswan, 500 miles south along the River Nile, up which they were ferried. Although very little of it remains today, a polished white limestone casing completed the construction, giving the pyramids a glorious, smooth, shiny appearance that must have dazzled under the African sun.

While some of the earliest chalk stone and wood enclosures surrounding the Stonehenge site in Wiltshire go back as far as 5,000 years ago, the famous standing stones were raised around the same time as the pyramids, between 4,400 and 4,200 years ago. The average weight of each of the outer stones is about 25 tons. The smaller 'bluestones' of the inner ring, each roughly 2 metres high and weighing 2 to 3 tons, are made of an igneous rock found around Presili, Carn Goedog and Pont Saeson in Pembrokeshire, South Wales, 150 miles away as the crow flies, and even further when transported by land and sea. Recent evidence suggests that these smaller, but still weighty, bluestones were cut from dolerite that had naturally cooled to form pillar-shaped slabs, making quarrying, if not their transport, easier.

Debates continue about how exactly these huge blocks were carried over such distances. The taller, larger, heavier slabs – the sarsen stones – that make up the outer ring, are made of silcrete (a silica-cemented sand and gravel). They were quarried relatively close to the henge itself. A recent study led by the physical geographer David Nash concluded that fifty of the fifty-two megaliths probably came from West Woods, near Marlborough, 15 miles to the north. Although sourced more locally, their greater weight still posed a formidable transportation challenge for the Celtic people whose vision and building prowess still astound us today.

By the time of the ancient Greeks and Romans, stonework was not only becoming more skilful, it was also capable of expressing great beauty. The Parthenon in Athens was completed around 432 BCE, while more than 500 years later the Pantheon in Rome made its magnificent appearance. The talents of these classical civilisations were equally impressive when they turned their hand to creating works of art. Carving statues out of massive blocks of marble – a metamorphosed limestone – created wonders such as the Venus de Milo and the 'discus thrower' of Myron.

Over a thousand years later, the architects and masons of the Catholic Church in Europe were building cathedrals of stone. These still take our breath away today. Pale, creamy limestones and fine-grained sandstones were generally preferred. These rocks were easy to cut and a joy to carve. The skills of the craftsmen created churches of soaring height, massive presence and great delicacy. Whenever possible, medieval builders quarried local rocks for their stone.

Durham Cathedral, now a World Heritage Site, was begun in 1093. It occupies a dramatic site on a promontory high above a bend in the River Wear. The architectural historian Nikolaus Pevsner described the cathedral's Norman Romanesque architecture as 'one of the great architectural experiences in Europe'. For my final undergraduate year at university, I lived on the top floor of Durham Castle's keep, overlooking the cathedral. I never tired of the building's massive splendour. The builders used local sandstones that were quarried in the city itself. The rocks were hewn from the Carboniferous Pennine Middle Coal Measures. These sandstones were laid down in rivers, lakes, deltas, estuaries and shallow seas around 310 to 318 million years ago when this part of Britain lay across the equator.

Sixty miles south, we find a similar story. Most of the beauty of York Minster that we see today was begun in 1230. The local stone found around these parts was a magnesium limestone from the Permian period. This was a time when the British Isles had drifted farther north towards the Tropic of Cancer. The rock was quarried near Tadcaster, just over 10 miles away to the south-west of the city of York.

However, as quality and ease of working were paramount, these medieval builders were not averse to transporting rocks quarried from places much further away. For example, the Normans, already familiar with working with the creamy-white Jurassic Limestones of Caen in Normandy, chose to import the French rock to build and face-stone their new cathedrals at Canterbury, begun in 1070, and Norwich, begun in 1096. Castles, too, used Caen stone, including parts of the Tower of London.

Cut stone is still used to face many civic and commercial buildings. Although more difficult to cut, polished granite, with its big shimmering crystals of feldspar, is still used to add a touch of weighty opulence to a building's lobbies and reception areas. One of Norman Foster's famous buildings, 30 St Mary Axe, London, better known as the 'Gherkin', has polished grey granite throughout to smarten its entrance and lift lobbies. If you want the headstone of your grave to weather the ravages of frost and rain, choose granite.

Rocks and stones in their raw state are also used more prosaically for everyday and practical purposes. The patchwork of drystone walls that tessellate hills and valleys are built using rocks hammered and chiselled from multiple, small, nearby quarries. Crushed limestones make a firm bed for motorways as they slide across the country on their sinuous way. Smashed and broken granites form the ballast on which railway sleepers and tracks lie and rest. And millions and millions of tons of rock aggregates – sands, gravels, pebbles - are excavated by the construction industry every year as it builds and bolsters, raises and fills.

Our theme in this book, however, is rock transformed by cunning, reduced by heat, changed by chemistry, processed with skill. Most rocks are made of a dense mass of closely packed, complex minerals whose crystals range from the minutely small to the beautifully large. In this sense, all rocks contain elements and minerals that potentially have some practical use. It is only when the sought-after element or mineral can be extracted from the rock with sufficient ease and economy that it becomes commercially worthwhile. In practice, this means that the cheaper the world price of a desired element or mineral, the higher must be its concentration in the rock to make its exploitation an economic proposition.

In the case of iron, let's say, concentrations of the recoverable metal from the ore might need to be as high as 25 to 30 per cent, depending on world market prices at the time. On the other hand, highly prized but very rare elements, such as platinum, need only be present in concentrations as low as a few grams per metric tonne, which is around five parts per million, to be worth mining.

However, all of this activity has consequences. We pay a heavy environmental price for this human ingenuity. In the process of making so much stuff, habitats are destroyed, oceans are polluted, climates change and biodiversity falls.

In December 2020, Emily Elhacham and her colleagues published a paper in the journal *Nature*. In it they describe how, year on year, humanity mines and makes more and more stuff. Concrete floors, steel-framed skyscrapers, tarmacked highways, brick houses, shopping malls, ships, cars, cups, bottles, smartphones, televisions and the vast infrastructure that surrounds our daily lives now weigh approximately 1.1 trillion metric tonnes. This is equal to or greater than the combined dry weight of all of the plants, animals, fungi, bacteria, archaea and protists on the planet.

Since 1900, human-made stuff has doubled in mass every 20 years. In 1900, artificial objects accounted for just 3 per cent of the world's biomass. Today, manufactured materials match nature's worldwide organic efforts. If current trends continue, anthropogenic mass will grow to three times the world's biomass by 2040.

Even more alarming, these calculations *don't include* all of the earth that is moved when ores are mined, coal is dug, roads are built, tunnels bored, foundations laid and seas dredged. They don't include all of the waste that manufacturing processes create along the way, all of the fuel burnt to power machines, all of the carbon dioxide released into the atmosphere.

Human beings have become a dominant, even dangerous, force on the planet. We are modifying the Earth to such an extent that we are creating a new geological epoch. We have named it the Anthropocene.

Degradation, pollution, global heating, climate change and the beginnings of the Anthropocene will be the subject of the book's final three chapters.

But first, let's concentrate.

Chapter 2

Concentrate

Before we introduce our chosen rocks and the stuff we derive from them, let's take a step back in time, a big step and a very long way back, and look at where all of the elements that make up the Earth, its minerals and crust came from in the first place. We'll start with the Big Bang.

Given that the universe continues to expand, cosmologists believe that if we extrapolate backwards, there was a time about 13.8 billion years ago when all of the matter and energy in the universe – radiation, planets, stars and galaxies – was concentrated at a single point at a density and temperature utterly unimaginable. For reasons that are still being theorised and debated, this singular point suddenly expanded in what the astronomer Fred Hoyle, never a fan of the idea, facetiously described as a 'big bang'. In spite of his dismissive intention, the name stuck.

After an initial phase of inflation followed by cooling of the Big Bang's primordial plasma, the first subatomic particles began to appear. Several hundred thousand years later still, and with further expansion and cooling, the first and lightest elements formed. Atoms of hydrogen predominated, but there was some helium, the second-lightest element, and a little lithium and beryllium.

Then, very slowly, over the next few hundred million years, gravity began to exert its influence. Hydrogen and helium atoms began to cluster to form vast gas clouds. Out of the clouds, these stellar nurseries, the first stars were born. For the first time, there was light in the universe.

Stars shine because the temperatures and pressures reached in their fiery hearts are so great that the lightest atoms, hydrogen and helium, begin to fuse to form heavier elements. In the fusion process, some mass is lost and converted into energy, including heat and light radiation. This is Einstein's famous equation, $E = mc^2$, in action. When stars can no longer sustain the processes of nuclear fusion, they reach the end of their lives. Many of the bigger stars end their days in one final explosive burst called a supernova. The stardust blasted into the cosmos as the supernova explodes contains many of the heavier elements forged deep within the star. These include oxygen, carbon, sodium, potassium, calcium, magnesium and lithium.

However, exploding supernovae don't quite explain the origin of the heaviest elements, such as lead, gold, mercury and uranium, and many of the rare earths. Even more exotically, when very massive stars catastrophically die, all that is left is a small, extremely dense ball of matter made of protons and electrons that fuse to create neutrons.

Typically, these 'dead' neutron stars measure a mere 20 to 25 kilometres across, but just one cubic centimetre of their mass would weigh a million metric tonnes. Unimaginable. And very occasionally, from time to time, somewhere in the vast universe two of these neutron stars collide. The explosive power is so great and the energy levels so high that astronomers believe this is when and where many of the heavier elements in the periodic table are first created.

These include gold and platinum. As the two colliding neutron stars spiral inwards, they also emit gravitational waves, again as predicted by Einstein.

The stardust from supernovae and neutron star collisions disperses to join the interstellar gas clouds of hydrogen and helium. Again, gravity exerts its inescapable pull on all of this thinly scattered matter. Slowly, slowly the dusty clouds of gas collapse forming new, *second-generation stars* such as our own sun. Surrounding these emerging newborn stars is a disc of dust and gas out of which planets begin to coalesce as gravity pulls matter together into larger and larger clumps. The heavier elements in the swirling disc condense and form the rocky planets closest to the star, which in the case of our own solar system include Mercury, Venus, Earth and Mars.

Most of the lighter elements in the solar disc remain as gases. They get blown further and further out by the solar winds to the point where they become sufficiently cold to eventually freeze and condense into solid grains. Gravity continues its effects, pulling the frozen gas grains together until they, too, begin to form planets – in our sun's case, the gas giants including Jupiter and Saturn. These are mainly composed of hydrogen and helium.

Now, the point of this space trip is to question how all of the familiar elements that we know on Earth managed to become sufficiently concentrated that they allow us to mine and refine, manufacture and manipulate them into all of the stuff that surrounds us in our everyday world.

Take iron, for example. Its abundance in the universe is only 0.11 per cent, but it forms 6.3 per cent of the Earth's crust. Even that isn't enough to make it economical to mine. Before it becomes commercially worthwhile, an iron ore needs to have an iron content of least 25 to 30 per cent.

Or take tin. Its abundance in the whole universe is even rarer. It exists in only four parts in every billion. Hardly there at all. However, in the Earth's crust it manages to creep up to about two or three parts per million – a thousand-fold increase in concentration. But before it becomes economic to mine, tin has to form about 0.4 per cent of the ore – that is, around 4,000 parts per million – another thousand-fold increase in concentration from its background presence in the crust.

We've seen how the formation of second-generation stars brings about a concentration of heavier elements in the form of the inner rocky planets. The next stage in the concentration process takes us back to the growth and evolution of the planets themselves. As they form, compact and coalesce, and as asteroids continue to rain down on them, for a while the rocky planets remain molten and hot. Like all molten masses, the heavier minerals and elements sink to the centre. In the case of the Earth, a core of iron and nickel developed.

The lighter elements of oxygen, silicon, aluminium, magnesium, potassium, sodium and calcium, along with some iron and tiny amounts of the remaining elements, rose to the planet's surface where they cooled and crystallised to form a solid crust. These elements rarely exist in their free state. They combine, mainly with oxygen, to form hundreds of different kinds of minerals. Given that oxygen and silicon are the two most abundant elements in the crust, these two unite to form various silicon oxides, which in their more complex guises give us a rich range of silicate minerals, out of which most of the crustal rocks are formed. These silicate minerals can range from simple quartz (silicon dioxide) to the ubiquitous, beautiful, shimmering granite feldspars, including potassium aluminium silicate (orthoclase) and sodium calcium aluminium silicate (the plagioclases).

There are, in fact, two types of planetary crust. The crust of the continents is thicker and generally made up of lighter elements and minerals. Oceanic crust is thinner and is composed of more 'basic', heavier elements and minerals. This is why the less dense continents ride higher above the Earth's mantle than the denser crust beneath the oceans.

The formation of the Earth and its fractionation into a dense core, a less dense mantle and a lighter crust therefore offers a second stage of mineral differentiation and concentration. We're beginning to see how elements that are extremely rare in the universe gradually become less and less rare as we follow the cosmological story from Big Bang to the formation and structure of the planets.

The penultimate stage in which particular minerals and their ores become sufficiently concentrated to make them worthwhile to mine involves geological processes. Plate tectonics, igneous intrusions, hydrothermal veins, volcanoes, metamorphism, weathering, erosion, evaporation and sedimentation all play a part as nature contrives to further differentiate and concentrate its elemental bounty.

The fourth and final stage of concentration introduces men and women into the story. We have learned to recognise, recover and refine a huge number of mineral ores to achieve 100 *per cent* concentration of all of the elements, including metals such as iron, copper and nickel. We have also managed to create entirely new materials out of rocks and their weathered remains, including concrete and glass.

Let's take just one example and appreciate how nature and humanity together go from 'hardly there at all' to 'pure metal' by quickly following chromium's cosmic journey. Chromium is a bright, silvery, corrosion-resistant metal. It's the metal that makes our bathroom taps shiny and, when added to iron, it makes the stainless steel for our cutlery.

After the formation and explosive deaths of the first big stars as spectacular supernovae, the abundance of chromium in the universe amounted to a mere 0.0015 per cent or 1.5 parts in every 100,000. When the disc of gas and dust around our sun eventually condensed to form the planets, and in the case of the Earth when the planet had finally cooled to form a core, mantle and crust, we find the level of chromium's abundance has increased a hundred-fold. In the Earth's crust, the average amount of chromium present is about 0.014 per cent or 1.4 parts in every 10,000.

The Earth and its crust are dynamic. Continents roam, tectonic plates dive, rocks heat and magmas melt. One result of all this igneous and metamorphic activity is that as the hot, molten rocks cool, and different metals and their minerals, each with its own freezing point and density, settle and separate out, they do so at different times and in different places. This results in metals and their minerals becoming concentrated in discrete layers and veins. In the case of chromium and its main mineral, chromite, ores become economic when the chromium content reaches at least 5 to 10 per cent. It is only in the final stage that we enter the story of elemental concentration. The chromite-rich ore is mined, roasted, leached, heated and chemically treated before pure metal chromium is obtained to make the taps shine and the cutlery gleam.

So that cosmic journey goes from 0.0015 per cent chromium in the universe to 100 per cent pure metal at the end of the refining process – all in four extraordinary stages.

* * *

In this book we shall follow the stories of a few key metals and minerals through the final two stages of these segregation and concentration processes: the geological processes and those processes devised by human ingenuity. Whatever the material, there is always a rocky tale behind its discovery, exploitation and manufacture.

A thin veneer of human endeavour and material manufacture coats much of the Earth's surface. Most of what we make and build is little more than transformed crust and rock. But what transformations! It is wondrous to think that when you look at a mountain or dig beneath the soil, somehow men and women have learned to turn all of this brute rock and fractured stone into palaces and planes, turbines and trains, bicycles and bridges, fridges and phones.

But once again, we need to remind ourselves of the downsides of all this cleverness. The very properties that have made our manufactured materials so successful have also created problems of environmental pollution and planetary distress. 'We are a clever and revolutionary species,' says the writer and broadcaster Richard Holloway, 'and we have spent our history achieving the impossible. But we have been bad at anticipating the effect of our inventiveness, not only on our own health and happiness but also on the health of the planet that is – still – our only viable home.' It seems that in the very act of transforming nature to serve our own ends, we have ended up changing the world and our relationship to it. There is more than a hint that humanity's hubris is now in danger of provoking both Gaia and Nemesis, the goddesses of the Earth and retribution, respectively.

Over the coming chapters we shall look at some of the most common and widespread of these everyday materials. We shall explore their nature, origins, extraction, manufacture and uses. Many of these we take so much for granted that we forget that behind each one lies a long, fascinating, often complicated sequence of events and inventions. The characteristics of many of these materials have also given rise to a rich culture of imagery and metaphor, design and architecture.

So, throughout these pages, let's take a look at iron and steel, plastics and glass, bricks and clay, aluminium and lithium, concrete and copper, rare earths and coal. Whenever you gaze up at a skyscraper, think iron and steel, concrete and glass. Whenever you use your mobile phone or switch on your computer, remember rare earths, lithium, copper, silver and gold. Whenever you fly in a plane, wonder about aluminium and titanium. Whenever you pour a drink from a plastic bottle or sit on a plastic garden chair, remember oil and its origins.

Chapter 3

Bricks, Pots and Ceramics

Bricks, pots and ceramics are among the earliest synthetic objects to be made by men and women. This ought not to surprise us. After all, on any rainy day, the ground beneath our feet is liable to become squelchy with mud. We can leave our footprints behind, and later, when the mud dries under a baking sun, that same slippery, soft stuff becomes firm and hard and there our footprints remain until the next heavy rains.

Some of the earliest evidence of our species' presence is the footprints baked into ancient river muds and shoreline sands dating back tens, even hundreds of thousands of years. In 2020, the chemical ecologist Matthew Stewart and colleagues reported their find of seven human footprints on the Arabian Peninsula. The prints had been baked into ancient shoreline lake deposits about 120,000 years ago. The discovery sheds an intriguing light on the spread of our species out of Africa.

Fire, too, was something with which our Palaeolithic ancestors had long been familiar. Fire kept us warm, cooked our food and helped ward off danger. Some 10,000 years or more ago, by putting two and two together, *twice*, we arrived at the basic idea that clay *plus* water *plus* moulding *plus* heat could produce wares made of earth – earthenware.

The basic ingredients for making pots and bricks are

therefore readily available and easily accessible. They are literally beneath our feet, not in the sense that all rocks are beneath our feet, but in the way that soils and clays, marls and muds will dirty our shoes and cling to our toes. Neither were these fluid, plastic, mouldable, almost magical properties of wet earths and clays lost on religious storytellers. A god fashioning the first man and woman out of clay crops up as a theme in many religions and ancient mythologies.

However, it did take a while to develop the technological skills needed to fashion, fire and harden clays into worthwhile and usable objects. One early example is the famous ceramic statuette known as the Venus of Dolní Vestonice. She is a small female nude, thought to be a fertility symbol, made of clay fired at relatively low temperatures. The little figurine was found in the Czech Republic and is estimated to be at least 27,000 years old.

One of the original uses of clays and soils was to make reed-framed or bamboo-woven walls windproof, if not always waterproof, by plastering them with wet earth. The result was a mud hut, although today most mud huts are made from air-dried, unfired bricks. The huts were built with a circular wall capped by a thatched roof. They still have a vital role to play in many of the drier parts of the world today, and if the mud is mixed with a bit of straw and bitumen, the unfired bricks even become waterproof and don't wash away in the rain. The Spanish word for 'mud brick' is *adobe*, and the word 'adobe' has come to mean any house or structure built of unfired mud bricks. Entire, ancient towns and villages have comprised adobe dwellings.

Wattle and daub houses were constructed on similar principles using cheap, local and readily available materials. As far back as Neolithic times, the technique involved building a wall of small timbers and woven wooden strips (wattle) to form a matrix that supported a mud-based daub. Depending on what is to hand, the daub can be made from materials as varied as wet clay, earth, lime, sand, dung, straw and animal hair. These ingredients have to be thoroughly mixed and then patted and pushed, slapped and slurped into the latticework and left to dry. The wall is then painted. Examples can still be seen in many of Britain's older timber-framed houses.

Cob walls are also constructed out of the surrounding soils and are still being built today, partly because of their environmental benefits. Cob is a mixture of sandy soil, clay and straw. Although clay remains a key ingredient, about three-quarters of the cob is made up of the sandy aggregate. The natural building and sustainability movement is a fan of cob constructions, and basic materials are locally sourced from the immediate environment. Cob houses are warm and dry, and they blend into the landscape out of which they have been fashioned.

* * *

The other thing to do with clay is to make pots out of it. It is difficult to know when people first began to make pots. Clay-shaped artefacts and remains have been found across Europe and Asia dating as far back as 27,000 years ago or more. Lighting fires on clay-based soils might have given people an early clue about what heat might do when applied to some types of earth. Further thoughts also suggest that, as basket-weaving preceded pottery-making, rather like wattle and daub houses, clay might have been used to plaster the basket and left to dry. Deliberate or accidental heating would then have left behind a clay pot vessel with its weave marks still decoratively visible.

It's also been suggested that it was not until men and women became farmers and began to live in permanent settlements that there would have been much point in making clay pots. Nomadic and hunter-gatherer lifestyles didn't lend themselves to shaping things out of clay. Carrying heavy, fragile pots around wouldn't have been a practical proposition. Pottery therefore didn't really take off until people settled, grew crops and adopted the farming life.

In his book *Pour Me: A Life*, the journalist and food writer A. A. Gill makes a wonderful claim for the humble pot:

> ...*the real quantum leap wasn't fire, it was the pot. Turning mud into utensils – that's what allowed us to boil food. That is the bubbling birth of cookery. When you put more than one ingredient into the pot it is a recipe.*

And in his book *The Body: A Guide for Occupants*, the writer Bill Bryson reminds us that cooking food broadens the range of what we can eat. It makes tough things tender, kills toxins, improves taste, boosts the number of calories released, reduces the amount of time we spend collecting food and so increases the amount of time for doing other things.

Pottery, whole or broken, has proved a blessing for archaeologists. Unlike many more perishable materials, bits of broken pottery can remain in the ground, relatively unaltered, for thousands of years. And as different cultures at different times in different places made pots that were peculiar to them, archaeologists have learned both to date and define people by the pots they made.

For example, the spread in the use of certain type of beautiful bell-shaped beakers across western Europe between 4,800 and 4,000 years ago inspired archaeologists to call the people who made the drinking vessels the Beaker Folk. In the Bronze Age, between 5,000 and 3,000 years ago, not only were people using pots for cooking, eating and drinking, but also for ceremonial purposes including funerals. Nearly every burial or cremation site found by archaeologists digging around in the Bronze Age contains pots, often highly decorated, either whole or in fragments.

The ceramics scientist Paul Rado rather wonderfully reminds us that pottery represents the perfect combination of what the ancient Greeks regarded as the four basic elements out of which the world is made: earth, water, air and fire. 'Pottery,' he says, 'is made of earth, shaped with water, dried in air, and made durable by fire.'

The essential feature of pottery manufacture is that the product is shaped in the cold state. When wet, clay is said to be 'plastic'; that is, it can be fashioned and moulded into any design we choose without losing its shape. We can even reshape it while it remains plastic and wet.

The word 'ceramic' is derived from the classical Greek for pottery, *keramikós*. In broad terms, ceramics refers to the manufacture of rigid clay products through shaping, moulding and heating. It is only when we put the fashioned and air-dried clay object into the flames that its chemistry changes, its structure turns hard, and its shape becomes permanent.

One of the earliest and simplest ways of making a pot was to roll the clay into long, thin strips. The coils are then woven and wound around to form a bowl or pot before being dried and fired. The still-wet coiled vessel might be smoothed using the hands to iron out the coiled ridges. However, it was the invention of the potter's wheel, possibly in the Middle East around 7,000 years ago, that marked the beginning of mechanising pot manufacture.

The process of making and baking pots has remained basically the same ever since. Of course, mechanisation and industrialisation have made the processes quicker, cheaper and more efficient, but pottery manufacturers still need clay to mould, air to dry and fire to harden before a pot, plate, jar, jug, vase, cup, saucer, washbasin or toilet bowl can emerge.

The technique of glazing – that is, coating with a substance that becomes glassy upon firing – has added to both the aesthetics and the durability of pottery products. As the science and technology of pottery-making have grown and developed, ever more sophisticated shapes and designs have become possible.

* * *

Being made of the same clay-based material, there is a parallel story to be told about bricks and their manufacture. We have briefly mentioned the use of mud bricks, which are made from moulded blocks of clay, air dried and then left in the sun to harden. In this process, the chemical and mineralogical composition of the clay brick has not undergone any fundamental change. If you put an unprotected air-dried mud brick in water, it will revert to wet clay. Nevertheless, in dry regions, hand-made, air-dried, sun-baked bricks are not only strong, they can last a long time. The 10,000-year-old Walls of Jericho, famously knocked down by Joshua and his horn-blowing priests, and now located in the Palestinian Territories, were built of such mud bricks.

But just as a fired clay pot hardens and takes on different chemical characteristics, so does a fired brick. Bricks, however, tend to differ from pots in terms of what is added, and in what quantities, to the basic clay mix.

The main ingredients introduced to the clay are usually sand or ground ash. These allow the moulded bricks to hold their shape. They also help reduce shrinkage on drying and lessen cracking when fired. A few loams and malms naturally have the right balance of clay, sand and chalk. This made them highly prized by traditional brickmakers. In practice, most clays are too sticky on their own to make good bricks, while ground-up shales need the addition of water as well as sand before they can be made plastic and suitable to be moulded into shape.

There are a number of ways to make a brick. The most traditional is known as the soft mud process. The clay-sand mix is made into a thick paste by adding water. It is then pushed and pressed into a wooden mould, or stock, before being tipped out, dried in the air, and then finally fired in a kiln. The process can be done by hand or machine. Bricks made by this method are called 'stock bricks'.

Many modern-day bricks are made from clays and shales that first have to be crushed and ground to a fine powder in huge machines. Fine sand, other materials and water are then added and mixed. Rather than shape the brick in a stock or mould, brick factories are just as likely to squeeze the stiff clay mix through a rectangular die to form long rectangular columns. A set of taut wires then descend across the strip, cutting off a series of bricks of the right size and shape.

The bricks are then dried before being fired in a long 'tunnel' kiln. These can reach lengths of up to 200 metres. As the bricks travel slowly through the tunnel, they pass through zones of increasing heat up to a maximum of 1,250°C, before finally being allowed to cool. They emerge chemically changed, partially vitrified, harder and often of a different colour from when they went into the kiln. The final colour usually depends on the type of clay used, although it is possible to add chemicals that under firing will produce a particular shade of red, orange, yellow, blue or grey.

Brick is a remarkably versatile building material. It lends itself to structures both large and small, domestic and commercial, industrial and artistic. Bricks have lined railway and canal tunnels. They have been used to build viaducts and aqueducts. Their durability and functionality are splendidly evidenced by the continued presence of, for example, the civil engineer Sir Joseph Bazalgette's vast network of London sewers, originally built of brick between 1858 and

1875 and still in use today. You only have to imagine what has flown along those underground tunnels for the past 150 years to be impressed by their brick linings' ability to withstand the cocktail of waste that has been flushed down them every day.

Bricks are also incredibly strong. In 1956, Albert Searle, an expert on clays and brickmaking, noted that the top side of one brick in a brick wall has a tiny surface area of a mere 0.25 square feet. This area has to support the weight and pressure of any further layers of brick built above. He calculated that it would take no fewer than 8,000 bricks to be laid above one base brick before it would crumble and collapse. He further observed that these 8,000 bricks, plus the mortar between them, would reach a height of 1,750 feet. That's an impressive load-bearing, crush-resisting statistic. All of which means that you can build surprisingly high using only bricks.

There are many lovely examples of monumental brick buildings, many using bricks of different colours. Sir Gilbert Scott's the Midland Grand Hotel at London St Pancras Station, which opened in 1873, uses 'polychromatic' brickwork to splendid effect. And if it's high you're after, you need look no further than Birmingham's Joseph Chamberlain Memorial Clock Tower, affectionately known as 'Old Joe', which rises well over 300 feet, making it the tallest free-standing clock tower in the world. It stands in the middle of the University of Birmingham campus and is built of Red Accrington brick, an iron-hard engineering brick. It is modelled on the Torre del Mangia campanile in Siena, completed in 1348.

There is some dispute about the origin of the term 'red-brick universities', with some believing it was first applied to the University of Liverpool's Victoria Building. But as the University of Birmingham was granted its university status in 1900, three years before Liverpool, it also claims the title, especially as it is the home of some splendid red-brick buildings, including Chancellor's Court and 'Old Joe' himself.

For many centuries, the colour of local bricks changed as you travelled from one part of the country to another. As transporting bricks was a cumbersome and costly business, it made sense to make bricks locally. As the colour of the clays and shales used to make the bricks varied geographically and geologically, so the bricks would leave the local kilns differently coloured. Although most clays share a common mineralogy, there are differences in the detail, and these differences give rise to clays that vary in colour from black to grey, from brown to orange, from purple to red, from yellow to ochre.

* * *

Some of the oldest geological clay beds and mudstones have been around for so long, and subjected to all kinds of tectonic ups and downs, pressures and heat, that they have been 'metamorphosed'. Crushed by huge forces deep beneath the Earth's crust, they turn into slates. These are fine-grained rocks that are 'fissile'; that is, they split into thin layers that develop in a direction perpendicular to that in which they experienced compression as they were dragged down beneath converging crustal plates. Slates can be quarried, crushed and powdered to form a clay mix for bricks, but they are much better split, shaped and used for roof tiles that sit prettily above the rich, soft tones of their sedimentary cousins, the fired bricks of clay.

For our purposes, we can define clays as very fine-grained, sedimentary rocks. They result from the weathering of older rocks, especially igneous rocks rich in feldspar minerals. These include potassium aluminium silicates known as orthoclase, and sodium-calcium aluminium silicates known as plagioclase.

In water-saturated environments, particularly those that are slightly acidic, rock minerals such as feldspars, micas and iron oxides break down and adopt new crystalline structures that are more stable in their cooler, less-pressured surface environments. These new 'clay' crystals are very small and go under the generic name of phyllosilicate minerals (*phyllo* from the Greek for leaf and thin-layered, which also gives us 'filo', as in the pastry!). These include montmorillonite and kaolinite. Given the cool, wet, rainy-acidic conditions necessary for their formation, most sedimentary clays therefore find themselves on, or relatively near, the surface of the Earth.

Once formed, the weathered grains and clay minerals can either accumulate where they originated or they can be washed away by rain and rivers, into the bottom of lakes, across flood plains or out to sea. Clays that accumulated at the bottom of lakes or on low-lying land are easy to dig. Before major industrialisation, most brickmakers or potters would site their workshops close to the source of their raw materials. Many of these disused clay pits have naturalised and become grassy hollows, ponds or lakes, which are a haven for wildlife. Many a Clay Lane and Brick Lane owes its name to long-abandoned clay pits and their nearby brick works.

Older clay deposits that have become buried beneath younger rocks are more likely to have become compressed, compacted and caught up within the normal crushes and pushes, ups and downs of the Earth's geological forces. Sometimes these clay beds will re-emerge at the surface as rock outcrops or they might be found as sedimentary layers some way beneath the surface.

Because clays represent the end product of the long-term weathering of older rocks, they can be found as bands in most geological periods, in all parts of the world, going back hundreds of millions of years. In Britain, workable brick clay and hardened mudstone beds can be found in the Carboniferous mudstones of northern England and among the Mercian Mudstone Group of the early Triassic period, running north to south through the Midlands.

However, we shall look at the story of just two particular clay deposits to illustrate how men and women continue to perform the wonderful art of turning soft, fine-grained rocks into hard bricks and delicate pots.

Running as an outcrop all the way north-east from the Dorset coastline through the Midlands and up to the banks of the Humber are rocks known as the Kellaways Formation and Oxford Clay Formation. They are ancient mudstones, siltstones and sandstones. These sedimentary beds formed approximately 156 to 165 million years ago in the late Jurassic period.

At this time, the British Isles were lying in the subtropics around latitude 30 degrees north. They were part of the vast Laurasian plate. Tectonic forces were gradually beginning to split this plate in two and by late Jurassic times the seas began once more to wash over what is now the British Isles. As the seas deepened, so the sediments being deposited became finer. A mud blanket began to cover the sea floor of what is now southern Britain. In the warm waters above the muds and clays lived ammonites, bivalves, cephalopods, fish and the occasional marine dinosaur, many of whose dead remains can now be found as fossils in the rock. It was in Jurassic rocks of this type that the remarkable pioneering palaeontologist Mary Anning of Lyme Regis found her hundreds of specimens. They included the famous sea-swimming dinosaur, ichthyosaurus, and in 1823 the first specimen ever found of a plesiosaur, another marine reptile.

Between what is now Oxford and the Humber, these clays and fine sediments would eventually consolidate, be tipped by tectonic forces to dip gently south-eastwards, and be named by twentieth-century geologists as the Oxford Clay. Being a soft rock, when exposed to the elements, the Oxford Clay easily weathers. It shows little resistance to the rain and wind and so forms flat, low-lying country, nicely seen in Bedfordshire and north to the fens of Cambridgeshire and Lincolnshire.

Today, many of the bricks manufactured in the United Kingdom use Oxford Clay. For example, as you approach Peterborough from the south by train you can see huge excavations and massive mechanical diggers that have dug deep beneath the surface, extracting the grey-black clays on an industrial scale. Super-tall chimneys, hundreds of feet high, pierce the wide skies as they take away the fumes from the brick kilns. Some of the older clay pits have now filled with water to form deep ponds. Wetland plants and water fowl are slowly adopting these watery acres as their new home.

One of the first manufacturers to exploit this Jurassic resource was the London Brick Company. In the late 1800s, the company discovered that the lower Oxford Clays and Shales around the village of Fletton near Peterborough had a high organic carbon content. This meant that after crushing and when fired, the bricks required less fuel to heat them. The end product became known as the 'fletton brick', which has been used extensively ever since to build houses, civic buildings and commercial properties throughout the south-east of England. In particular, the fletton brick was the brick of choice for the hundreds of thousands of houses being built in London from late Victorian times onwards.

Perhaps the company's most notable marketing coup was to win an order for ten million fletton bricks to help build the Catholic Church's Westminster Cathedral in central London, whose Byzantine Revival style of architecture looks just as fresh and lovely today as it did when it first opened in 1903.

* * *

In complete contrast to the rough, abrasive feel of a brick is the translucent beauty of a piece of delicate porcelain pottery. Yet both are created from heating clays, along with the judicious addition of a few fine aggregates, in kilns that can be fired to temperatures in excess of 1,000°C.

Now, as far as potters (and indeed) brickmakers are concerned, this is the interesting bit: all clay minerals have an affinity for water. This means that in the presence of water they become 'hydrated'. Water becomes incorporated into their chemical structure and crystalline make-up. Some clay minerals can swell to twice their thickness when wet. It is these chemical and physical properties of clay that make them plastic, mouldable and squidgy when wet and therefore ideal for shaping pots and fashioning bricks.

It is only when they are heated to high temperatures that clays and clay minerals undergo a number of crucial, irreversible chemical changes. The clay goes from a soft, sensuous plastic body to one that is rock-hard, durable and impervious to water; from a weak, easily deformed material to a solid, obdurate mass; from having feet of clay to hitting a brick wall.

In the initial drying process and the early stages of heating, the moulded clay loses most of the atmospheric water trapped between its tiny particles. Most clays contain some carbon, organic materials and sulphur. These burn off at between 300°C and 800°C. At these temperatures, further dehydration takes place as water that had chemically combined with the cool clay minerals is lost.

However, it is only when kilns reach even higher temperatures that we begin to see the clay and quartz minerals undergo fundamental changes and adopt new chemical configurations and crystalline structures. Clay particles begin to stick to one another in a process known as sinter-ing. Some melting also takes place, and vitrification, filling in the gaps between particles. This increases the fired clay's strength and hardness. From the original clay minerals, new aluminium silicate minerals appear, including cristobalite, spinel and the needle-shaped mullite, first found naturally on, and named after, the volcanic island of Mull.

Unlike the original clay minerals such as montmoril-lonite and kaolinite, these new fire-formed minerals do not have an affinity for water. You can leave a fired stoneware pot in water and it will not revert to the soft, plastic clay from which it was born. Indeed, in general, the higher the firing temperature, the more durable and water-resistant the pot becomes. At the top of this firing temperature range are the porcelains and bone china. For them we need particularly fine clays. These are the china clays. Their genesis contrasts with the muddy marine origins of the Jurassic clays that provide so many brick factories with their raw material.

The china clays are the purest of clays. They result from the surface weathering of granites. Geologically speak-ing, china clays are among the youngest of deposits. They remain *in situ*, at the site of their formation. China clay, or *kaolin*, forms mainly from the breakdown by rain and water of the feldspar minerals that are present in such abundance in granites. Weathering can penetrate the upper layers of a granite down several hundred feet, leaving behind the dis-tinctive white kaolinite clays.

The term 'kaolin' derives from the name of a hilly ridge in Jiangxi Province, south-eastern China, from where the clay was first obtained. In the seventh and eighth centuries, the Chinese developed techniques for using kaolin to make beautiful white porcelain pottery, which became known as bone china. China clays are found in many parts of the world wherever there are granites that have been exposed to water and rain.

However, it wasn't until the 1740s that china clays were discovered on the weathered fringes of the granites of Bodmin Moor, just north of St Austell in south-west England. These molten granites, as well as those of Dartmoor and Exmoor, first up-welled about 290 million years ago beneath a once-great mountain chain known as the Variscan Mountain Belt, which ran east to west. It is only the aeons of time, the continued play of geological forces and the relentless processes of erosion that have brought the upper reaches of these ancient granite intrusions to the surface.

Although more recent extractions of kaolin have involved digging and the use of 'dry' mining methods, a long-established technique has been to blast the clay-rich ground with high-pressure hoses. This method has the advantage of leaving many of the impurities behind as the water washes down, forming a fine clay slurry that is collected in reservoirs. After further refining, the clay suspension is run into settling tanks, where the surplus water is drawn off. A final process of filtering and drying leaves the china clay ready for transport and manufacture. The very low iron content makes these clays not only white but translucent when fired.

There are many grades of pottery. Quality generally increases with the purity of the original clay body and the temperature at which it is fired.

Low-grade porous pots include common pottery, majolica and earthenware. Dense, non-porous pottery includes stoneware, vitreous china and soft porcelain.

At the top of the range, we have the hard porcelains and bone china. In practice, most of us, when we bring out 'the best china', are talking generically about the vitreous ceramics: the white, translucent porcelains. 'China', of course, is the simple contraction of 'chinaware' to describe the porcelains that were originally imported from China. They are as fragile as they are beautiful. Bulls should not be let anywhere near them, especially in shops.

Less glamorous is the etymology of the word 'porcelain' itself. *Porcella*, in Italian, means 'little pig', and is often used to describe a small sea-snail whose shell is white and translucent. The shell looks rather like the delicate pottery to which it gives its name.

The classical composition of hard porcelain is 50 per cent fine clay, 25 per cent feldspar and 25 per cent quartz. After a low-temperature hardening, the pot is dipped into a glaze before being fired at a higher temperature of up to 1,400°C. Glazing is the process of covering the pottery with a thin layer of specialised glasses. Once the pot has been dipped in the glaze, or 'slip', it is fired. The high temperatures 'melt' the glaze, which then covers the pot in a glassy surface. Glazing not only adds to the colour range and the aesthetics of pottery, it makes it smooth to the touch, non-porous, more resistant to chemical attack and mechanically stronger.

Bone china, as the name implies, contains powdered calcined bone ash, including the shinbones of cows and horses. It was Josiah Spode who, in 1789, first established the ideal composition of bone china: 50 per cent bone, 25 per cent Cornish stone (made of feldspar and quartz) and 25 per cent china clay. However, it was Spode's contemporary, Josiah Wedgwood (1730–1795), who led the way in turning pottery from a craft into a major industry. He came from a family of Staffordshire potters.

Originally based in Burslem, one of the Potteries' 'five towns', Wedgwood pioneered the practice of getting his workforce to become task specialists rather than all-round potters. This represented an early form of the division of labour based on particular people carrying out specific tasks. There were a fearsome number of stages involved in turning lumps of clay into functional pottery. The industry, to this day, is relatively labour-intensive and still requires some highly skilled men and women to monitor the drying and firing, and to decorate the wares. With his drive and flair, Wedgwood helped spearhead the Industrial Revolution, achieving great wealth in the process. From building canals to get his clays to the factories, to following the latest developments in 'natural philosophy', he was a key figure in moving the Age of Reason into the Age of Technology.

His entrepreneurial drive led him to think scientifically and experimentally about how to make pots that were cheaper, better glazed and more beautiful. His business nous and marketing flair also saw him expanding his client base until eventually he achieved a worldwide market. In 1765 he opened a showroom in London. After he won an order for a gold-leaf porcelain tea service from Queen Charlotte, wife of King George III, there was no looking back.

But Wedgwood was also ahead of his time in giving thought and showing concern for the living standards of his workforce. Much of this stemmed from his Unitarian, 'dissenting' upbringing, the strongest expression of which was seen when he took up a staunch position as an anti-slavery campaigner. The Wedgwood family tree is quite remarkable. Charles Darwin, he of the theory of evolution fame, is Josiah's grandson, while a couple of generations after that, the composer Ralph Vaughan Williams can trace his ancestry back to both his great-great-grandfather, Josiah Wedgwood, and Charles Darwin, his great uncle. How's that for a dynasty?

* * *

Moulding bricks and making pots are among the oldest of the crafts that have learned to turn earth and rock into objects of great practical value. It is easy to take them for granted, so commonplace are they. Gaze across any urban, and more particularly any suburban, landscape and you will see mile after mile of things built out of brick. Houses of brick. Walls of brick. Brick bridges and converted Victorian factories built of brick. Driveways paved with brick.

Each day we drink out of cups, eat off of plates, serve food on dishes, and place fruit in bowls. This is the crockery of everyday use, the pottery of the kitchen. There are toilet bowls and sinks, the ceramics of the bathroom.

There is another kind of pottery, the pottery of ornament and art, fashioned in studios and sold in galleries. Ceramics are found in laboratories. They resist the corrosive power of chemicals. They are good insulators of both heat and electricity. Dentists can even make you a tooth out of a dental ceramic. They can brighten your smile with a porcelain veneer. Next time you are faced with the dazzling beam of a set of unbelievably white, perfect teeth, think weathered granite, residual rock and claggy clay. That might wipe the smile off their face.

Chapter 4

Copper

Aluminium makes up about 8 per cent of the Earth's crust; iron around 6 per cent. Copper, though, is much rarer. Its rate of crustal abundance is a mere 0.0068 per cent, just one-thousandth that of iron. This fraction is not dissimilar to some other 'base metals' with which we are familiar, including lead (0.001 per cent), zinc (0.0078 per cent) and tin (0.00022 per cent).

To understand how copper minerals can be mined economically, we need to look at how they become concentrated in rock bodies to make their exploitation worthwhile. But first, a few physical facts about copper and a bit of history.

Native copper is unusual among the metals in that it is brightly coloured – a shiny coppery red! Most metals are white or a light silvery-grey in colour. Think tin, iron, silver, lead, mercury, aluminium and nickel, to name a few. Pure copper metal does occasionally occur naturally. This is probably how it was first found and appreciated by Neolithic people. They soon learned to work the soft metal, beating and hammering it into beads, ornaments and tools.

However, its widespread use back in the late Neolithic period and Bronze Age, around 6,000 to 7,000 years ago, was based on extracting the metal by smelting rocks. For this to work, the rocks to be smelted needed to have reasonable concentrations of one or more of copper's many minerals.

Neolithic men and women were probably first alerted to the effects of heat on certain minerals by accident and observation. Metals with low melting points such as lead (327°C) and tin (232°C) were the first to reveal their presence when their minerals were heated. Nevertheless, fierce fires built on particular soils and rocky surfaces might have also yielded molten beads of bright, shiny-red copper. Again, putting two and two together, along with some trial-and-error experimentation, eventually gave rise to a basic understanding of the principles of smelting rock ores to release their pure metals. Fire, often regarded as the destroying element, became the transforming element.

Jacob Bronowski writes in his book *The Ascent of Man* that the great transformation that helped civilisations leap from the Stone Age to the Age of Metals to the Age of Enlightenment and the scientific revolution was made possible by fire:

> ...*it is the use of fire to disclose a wholly new class of materials, the metals. This is one of the grand technical steps, a stride in the ascent of man, which ranks with the invention of stone tools; for it was made by discovering in fire a subtler tool for taking matter apart. Physics is the knife that cuts into the grain of nature; fire, the flaming sword, is the knife that cuts below the visible structure, into the stone.*

Rather like the parent metal itself, copper minerals tend to be vibrantly coloured. Azurite is a rich, deep blue;

malachite an intense green; chalcopyrite a brassy yellow. This means that copper minerals were easy to spot and hence they became early candidates for the heat treatment.

Various methods were developed to mine copper mineral-rich rocks. As well as brute force using stone axes and bone tools, more subtle techniques such as lighting fires beneath the rocks and then quenching them with water, causing them to crack and fracture, were developed.

Around 5,000 to 6,000 years ago in several parts of the world, including the Mediterranean and Middle East, smelting technologies became increasingly sophisticated. The crushed ore was placed inside a brick, clay-lined furnace. Hotter than burning wood, charcoal began to be used as a fuel. Copper's melting point is 1,084°C, which meant that it became necessary to find other ways of raising and maintaining the temperature. One method was to introduce a strong up-draught. Narrow openings facing the winds provided one solution. The use of bellows was another. Small chimneys also helped. Methods then had to be developed to collect the molten metal so that it could be poured into moulds.

In ancient scripts, various names had been given to the metal. The ancient Greeks called the island of Cyprus 'Kyprios'. Copper was found and mined on the island and so the Romans, following the Greek, named the metal *cuprum*, a contraction of *cuprium aea*, the 'Cyprian metal'. Picking up from the Roman Latin, the Proto-Germanic languages of Northern Europe named the metal *kupar*, which later became *coper* in Old English, before finally arriving at today's word, 'copper'.

However, although coppersmiths soon learned to hammer copper into sheets and fashion it into tools, bangles and other ornaments, its malleability and softness meant that its practical uses were limited. After a few centuries in which Neolithic metallurgists continued to gain in expertise, it was discovered that mixing molten copper with other molten metals produced seemingly new materials with very different properties. These new hybrid metals we now know as alloys. Copper has quite a few to its name, most notably brass (copper and zinc) and bronze (copper and tin). Easily melted with most metals, today there are well over 500 known alloys of copper. Bronze, of course, gave its name to nearly 2,000 years of human history. Stretching roughly between 5,000 and 3,200 years ago, the Bronze Age was sandwiched between two other rock-based technologies: the earlier Stone Age and the later Iron Age.

Bronze is harder and more durable than either copper or tin alone. These were attractive properties to early metalworkers. The cutting edges of bronze swords, axes and other tools were more effective than copper on its own. However, when metalworkers eventually learned how to smelt iron and fashion even sharper weapons, stronger tools and more robust agricultural implements, the age of bronze was over. Although copper, bronze and brass would continue to be used for pots and bowls, jewellery and decoration, baubles and bells, coins and statues, it would be another two and a half thousand years before a new and even more exciting property of copper would be discovered.

* * *

From the outset, it was recognised that copper and its alloys were good at resisting corrosion. When exposed to oxygen, carbon dioxide and moisture in the air, copper does react, but the chemical changes go only skin deep. The bluish-green patina, or *verdigris*, that forms on the surface of copper-lined roofs or copper-cast statues, acts as a protective barrier, preventing further corrosion. Depending on such things as humidity and air pollution, the exact chemical composition of the patina will vary, but most are made up of complex copper compounds, including copper oxides and carbonates, many of them hydrated. The characteristic blue-green patina of copper-lined church roofs, the pale greens of the Statue of Liberty and the dome of Berlin Cathedral add aesthetic appeal as well as resistance to corrosion.

Copper plating was also used in the eighteenth century to line the bottom of wooden-hulled ships, which helped to prevent them from rotting. Shipworms and barnacles that would otherwise attach to and attack the vessel's timbers were put off by the presence of the metal. Over the years, this practice gave rise to the phrase 'copper-bottomed'. Any enterprise, plan or deal that was described as 'copper-bottomed' was guaranteed to be safe, reliable, trustworthy and without holes. It would hold water. It certainly would not sink.

However, although copper is generally found to be a metal that doesn't react readily with many other compounds, it does dissolve in ferric chloride and some acids, including nitric and sulphuric acids. This chemical feature eventually gave rise to a new art form.

In the late fifteenth and early sixteenth centuries, craftsmen and craftswomen developed the technique of etching designs onto various metal plates, including copper. A copper plate would be coated with a thin, resistant layer of wax. A design, picture or portrait would then be 'etched' onto the wax, revealing the bare metal beneath. The plate would be immersed in a bath of acid, which would eat into the exposed metal, leaving the waxed remainder untouched.

After the wax layer was removed, the artist would be left with a copper plate on which the design, picture or portrait would be visible as an etching. The plate was inked over and then wiped, leaving ink present only in the etched grooves. The plate was placed in a press together with a sheet of paper. The pressed paper picked up the ink from the etched lines. The result was a print of the original drawing. The process can be repeated over and over again, giving the artist multiple copies of his or her print. This made etchings potentially more of a money-spinner than a one-off oil painting.

Many artists have made, and continue to make, prints using etching as their method. Today, etchers prefer to use ferric chloride as their 'mordant' (fixing agent) to react with, and mark, the copper plate. But few will surpass Rembrandt in their etching skills. Rembrandt remains an outstanding exponent of the art and craft of making prints from etched plates. His self-portrait of 1630 is a particularly powerful example of the technique, with its simple yet dramatic form, achieved as the artist explored his own face and the intense passions it could display.

* * *

Copper's ability to conduct heat was recognised very early on in its manufacture and development. Copper pots and pans became commonplace in kitchens and fire hearths. However, it was the discovery of electricity and copper's ability to conduct an electric current with very low resistance that gave the metal a whole new lease of life.

It had long been known that certain fish could give anyone who touched them a tingly shock, or that feathers and hair would be attracted to an amber rod that had been rubbed with fur, or that lightning was something around which interesting physical things happened. However, it would not be until the seventeenth and eighteenth centuries that 'natural philosophers' began to investigate these phenomena more scientifically. Work by people such as Luigi Galvani, Benjamin Franklin and Alessandro Volta gradually established that electricity was some kind of basic force of nature.

Early pioneers also learned how to generate an electric current at will by inventing machines based on the idea that friction between certain materials generated an electric charge. This paved the way for the development of 'electrostatic machines'. These were hand-cranked devices that used friction to generate short bursts of electricity. Many of us who studied science at school will have been given an electric shock delivered by a Wimshurst machine being rotated with enthusiasm by a smiling physics teacher. Similar sadistic effects could also be achieved by getting students to play with a Van de Graaff generator. Happy days.

A fundamental breakthrough in the generation of electricity was achieved by Michael Faraday during the 1820s and 1830s while he was working at the Royal Institute in London. Faraday, the son of a blacksmith, had been impressed by Hans Christian Ørsted's discovery that a magnetic needle in a compass was deflected when placed near a wire with an electric current flowing through it. So, wondered Faraday, if an electric current can produce a magnetic field and deflect a compass needle, perhaps the reverse is also true: that a magnetic field can create an electric current. He began experimenting.

Copper was already known to be a very good conductor of electricity. Indeed, among pure metals at room temperature, after silver, copper has the second-highest electrical and thermal conductivity. What Faraday discovered was that by moving a magnet inside a tube of coiled copper wire or, equally effective, by rotating a copper disc between the poles of a magnet, he could induce an electric current in the copper. The phenomenon is known as electromagnetic induction. Faraday showed that *mechanical action* – the rotation of copper wires inside a magnetic field – *could be converted into electrical energy*. The Age of Electricity was born. The innovation was destined to have a profound impact on the development of modern civilisation and its utter dependence on reliable systems of electricity generation, transmission and conversion.

One more insight into the relationship between electricity and magnetism helped make sense of the phenomenon, not just practically but also theoretically, and in the process reveal one of nature's fundamental forces.

James Clerk Maxwell was a Scottish mathematician, physicist and, let's face it, genius. In 1861, using equations that were both profound and revelatory, he showed that *electricity and magnetism were in fact manifestations of the same thing, the same force.* When charged particles are accelerated by an electric field, their movement produces an oscillating wave of electric and magnetic radiation. The wave travels through space at the speed of light. This coincidence helped Maxwell realise that light itself was, in fact, an example of electromagnetic radiation with a particular wavelength and frequency.

Speculation soon led to other examples of electromagnetic radiation being discovered, each with its own wavelength and frequency, but again travelling at the speed of light in a vacuum. In fact, electromagnetic radiation exists across a wide, continuous spectrum of wavelengths. They range from those with very long wavelengths, such as radio waves, through those with intermediate values, including microwaves, visible and ultraviolet light, to very short-wave radiations known as X-rays and gamma rays.

These waves also carry energy – electromagnetic energy – and can therefore carry information, too. These insights opened up the possibility of communicating at the speed of light by electronic means. Whole new industries and inventions quickly emerged based on electricity and magnetism, radio waves and microwaves, visible light rays and infrared rays, X-rays and gamma rays. As men and women learned ever more clever ways of using this force of nature, somewhere along the technological line there would be copper playing a vital role in helping first to generate, and then to conduct, these forms of energy from one place to another.

The discovery of the link between magnetism and electricity has become the basis on which most electricity is produced, even to this day. Electricity generators are just very large versions of Faraday's original idea of 'armatures' of looped coils of copper wire spinning between very big magnets. It is also possible to have the coiled-copper armature remain stationary while powerful magnets on the outside rotate rapidly around it.

In modern-day generators, the energy to rotate the armatures can come from a variety of sources. One of the first methods was to heat water by burning coal. The steam produced is used to drive a turbine that spins the copper coil inside a casing of powerful magnets. So, rather wonderfully, in a series of steps and energy conversions, the *thermal energy* locked up inside the coal is released to boil the water to make the steam, which is converted into the *kinetic energy* of the rotating coil; this in turn generates *electrical energy*. Oil, gas and nuclear fuel can also be used to produce steam to drive the generators. Heat, motion, electricity.

It is also possible to generate electrical energy directly from kinetic energy. This is the principle on which wind turbines, or 'wind energy converters', work. The wind – kinetic energy – turns blades, behind which sits the electricity generator. Or you can use water and the force of gravity. In the case of hydroelectric power, it is the force of flowing water downhill that turns the turbines. You can also use the gravitational pull of the moon to create electricity. As ocean tides ebb and flow, it is possible to direct the huge volumes of water through submarine turbines and their generators. Tidal energy. Moon-powered electricity. Now isn't that a romantic idea, as well as an environmentally friendly one?

But whatever energy conversion strategy you use, copper wire lies at the heart of the generating process. Aluminium, because of its relative lightness and cheapness, is the metal generally used to build the heavy-duty power lines that take the electric current away from the power stations and wind turbines, feeding power into the national grid. However, copper wiring comes back into its own once the power reaches the consumer. Most medium-sized properties, businesses and domestic premises use copper wiring as their primary conductor for switching on lights, connecting to television sets, plugging in kettles, and powering all of the electrical gadgets that crowd our lives.

The gadgets themselves will more than likely have copper connections and copper circuits running through their electrical innards. It is impossible to imagine the modern world without electrical machinery, devices and instruments. Everything from the electric motors that start our cars and tumble laundry in our washing machines to the electrical circuits that weave their way inside and around our radios, music players, phones and computers involve copper somewhere in their make-up. Taking all of these electrical goods and circuits together, along with copper piping and plumbing, a typical home will contain anything up to a hundred kilograms of copper.

* * *

So, it's no surprise then to learn that the world produces around 20 million metric tonnes of copper a year. Of this total, around 40 per cent is used in the electronics and electrical goods industry, 30 per cent in building and construction, 10 per cent in industrial machinery and equipment, and 10 per cent in transportation systems.

Small amounts of copper are added to make a variety of copper compounds. These occur in pesticides and some medicines, and as chemicals to control fungal growth on grapes and algal growth in outdoor swimming pools.

All of which means that an awful lot of copper ore needs to be mined each year to feed the world's addiction to, and dependency on, the metal.

* * *

Copper combines with many other elements to form more than 150 different copper minerals. However, only a few of these occur in high enough concentrations to make mining economically worthwhile. In nature, copper minerals form in one of three basic ways.

Some result from igneous and volcanic activity. As hot magmas cool deep below the Earth's crust, different minerals begin to crystallise at different temperatures, with some appearing and concentrating earlier than others. Many metallic minerals, including those of copper, also dissolve in the hot fluids that push and flush their way up into fissures and faults above the rising magma. When the hot, chemically saturated waters eventually cool and solidify, they form mineral-rich veins. One of the most common copper minerals to be deposited as a result of such hydrothermal activity is *chalcopyrite*, with its brassy, gold colour. It is chemically described as copper iron sulphide ($CuFeS_2$).

Other copper minerals form when pre-existing copper-rich rocks weather in the air and rain. The various copper oxides that result include dark-red crystals of *cuprite* (Cu_2O) and the bright, distinctive green mineral *malachite* ($Cu_2CO_3(OH)_2$).

A third group forms as igneous and hydrothermally deposited copper sulphides are leached. As air and water percolate through coppery rocks, other minerals are dissolved away, leaving behind copper-enriched minerals, including grey-black crystals of *chalcocite* (Cu_2S). The sedimentary rocks of the Central African Copperbelt contain vast reserves of such disseminated copper minerals. The host rocks were originally deposited in shallow, sinking marine basins more than 800 million years ago. Copper deposition occurred gradually over the next two or three hundred million years as the sediments were buried deeper and deeper. Mineral-rich brines that had leached from basement rocks flowed through the sedimentary mudstones, sandstones and limestones precipitating the copper sulphide minerals, along with those of cobalt, lead, silver and zinc.

In these various ways, copper ores can occur in a wide variety of geological environments, from primary igneous activity to secondary, sedimentary processes. In many cases, the vibrant colour of copper minerals is a giveaway of the metal's presence. We began this book with a visit to Alderley Edge. Nearly 4,000 years ago, in the Bronze Age, veins of malachite were found in these 250 million-year-old Triassic sandstones. The veins were formed by a series of complex processes involving leaching and brine-rich waters circulating in the cracks and fissures of the consolidating desert sandstones. Bronze Age miners hacked deep shafts into the rock. This makes them some of the oldest copper workings to have been found in Britain.

For a brief time in the late eighteenth century, the Parys Mountain mine in Anglesey, Wales, was the world's leading producer of copper ore. Although the ore was low-grade in quality, it was present in large quantities, near the surface and easy to mine. Mineralisation in this area is associated with highly altered volcanic rocks and mudstones. As the price of copper fluctuates on world markets, recent surveys wax and wane in their assessments of whether or not it is worth reviving these old Welsh mines.

In 2019, the world's leading copper producers in order of output were Chile, Peru, China, the Democratic Republic of the Congo and its famous Copperbelt, the USA and Australia. The copper mines of the top two producers stretch along the length of the Andes Mountain Range on the eastern shores of the Pacific Ocean. This gives us a strong clue as to the type and likely origin of these vast ore reserves. Geologists describe them as *porphyry copper deposits*. These types of deposit currently account for up to 60 per cent of global copper production. In order to explain their occurrence and extent, we have to consider how mountains are built and how the plates that make up the Earth's crust get shunted around the planet.

Beneath the Earth's crust, rocks of the mantle are hot enough to flow, ever so slowly. They are said to behave like a plastic. The hot rocks of the mantle set up vast, slow-motion convection currents that rise, cool and then sink, only to be reheated and rise again. The tops of these convection cells drag the lighter Earth's crust along with them.

When the rising currents of two adjacent cells reach the top of the mantle, they *diverge* and pull apart the crust above them. This creates a 'rift' in the crust along which hot magmas can rise and erupt to form volcanoes, new basaltic crust and ridges thousands of miles long. The Mid-Atlantic Ridge with its many volcanic islands is one example. The African Rift Valley is another. As the diverging currents pull the crust apart, they also open up new seas and widening oceans.

In contrast, when two adjacent cooling mantle currents *converge* and sink, they drag opposing blocks of crust, or 'plates', towards each other. This results in a planetary crash site in which two mighty crustal plates plough into one another. The denser oceanic crust of one advancing plate tends to be thrust beneath the less dense, more buoyant continental crust of the other. In the collision process, crustal rocks along these converging margins get pushed up into great mountain chains.

The Alps, even today, continue to buckle and bend as the African plate remorselessly pushes into the European plate. The Indian plate is another chunk of crust grinding north, riding on the slow-flowing mantle currents beneath. As India drives into and beneath the Eurasian plate, the crustal rocks have been, and continue to be, pushed high, in this case to form the Himalayas. And in South America, the Nazca plate in the eastern Pacific is slowly but inexorably, at the rate of about 3.7 centimetres per year, ploughing into and plunging under the South American plate, pushing up the Andes in the process.

In these slow-motion crustal crash zones, one plate usually dives beneath the other in a process known as *subduction*. The crustal rocks that are 'subducted' beneath the rising mountain chain become heated as they thrust down into the hot upper mantle. That's where they begin to melt.

As huge plumes of these molten rocks and mineral-rich, hydrothermal fluids rise, flow, fracture and force their way into the fissured rocks of the mountains rising above, they cool and form mineral-rich veins and metalliferous igneous intrusions. However, different minerals precipitate out and begin to grow at different temperatures in different places. This means that as the superheated waters rise, spread and cool, different metallic minerals get deposited and concentrated in different zones around and above the intrusion. These are known as *disseminated ore deposits*. Above these igneous intrusions, hundreds of mineralised veinlets finger their way into the surrounding geology. As well as copper minerals, these porphyritic zones are often rich in other metals too, including molybdenum, tin, rhenium and sometimes small amounts of silver and gold.

Some of the largest, most productive copper mines in the world can be found in northern Chile, on the western slopes of the Andes Mountain Range. These copper porphyry deposits formed between 50 and 100 or so million years ago in the subduction zone where the Pacific Nazca plate began its dive beneath the western edges of the rising South American plate. Although the actual concentrations of copper are quite low, often less than 1 per cent at best, the presence of the ores so near the surface and in such huge quantities makes their exploitation economically worthwhile.

The first stage in processing the copper ore usually takes place close to the area in which it is being mined. After the ore is crushed and ground, copper concentrations of up to 30 per cent can be achieved through various physical and chemical processes. Copper sulphide ores are sent to smelters before a final stage of electrolytic refining yields pure copper metal. Copper oxide ores are subjected to a number of chemical treatments before pure copper can be gathered by processes known as 'electrowinning' and 'electroextraction'. At the end of both processes, the pure copper metal is ready to be used in the manufacture of electrical goods, electrical machines, the construction industry and transportation systems.

China, Chile, Japan, the USA and Russia are currently the world's leading producers of smelted and refined copper. As the world's leading importer of copper ore and producer of the pure metal, China is also far and away the biggest consumer. China accounts for nearly 50 per cent of the world's annual consumption of copper, with Europe, the USA and Japan following a long way behind.

Copper is one of the most traded metals in the world. It underpins so much of our modern life. The circuits that weave and thread their way through our electrical devices and connect us to everything and everyone depend on copper. Copper is key to the digital age in which we all now live. The need for copper continues to grow as we depend more and more on information technology and the devices that support it. The global push to reduce carbon emissions by travelling in electrically powered vehicles only increases the demand for copper further. But there is also an irony. The less we pollute the air by going electric, the more we degrade the land as we mine it for copper.

This ever-growing demand for copper means that it is not cheap. Depending on the state of the world economy, copper ranges in price from £30 to £50 a kilogram. A solid copper sphere, not much bigger than a snooker ball, weighs about a kilogram.

* * *

As with all major mining and industrial activities, producing pure copper requires a great deal of energy. The ultimate source of this energy determines the 'greenness' of the production process. Heat and electricity derived from burning fossil fuels, including coal and oil, clearly add more of the greenhouse gas carbon dioxide to the atmosphere, which is a bad thing. In contrast, electricity generated by wind, water and sun is much 'cleaner' and much preferred.

The smelting process itself can create pollution. Large volumes of low concentrations of sulphur dioxide can be emitted. When sulphur dioxide enters the atmosphere, it readily combines with water molecules to form weak, but potent, vapours of sulphuric acid. When these vapours condense, they fall as acid rain. Such rains can damage crops, defoliate trees and dissolve limestone, whether in its natural state as a rock or as a building stone used to construct churches, grand houses and municipal offices.

Land degradation is also a problem. There is no getting away from the fact that as the amount of copper, even in the richest ore deposits, is relatively low, removal of enormous amounts of soil and rock are necessary to make mining economically profitable. Huge open-cast pits degrade the landscape for miles around. Forests are felled. Grasslands are stripped bare. Animals lose their natural habitat. Unless regulations are made and upheld, the ore and the chemicals used to purify it can escape and pollute rivers and lakes, killing fish, birds and other wildlife for miles around.

On the plus side, copper is endlessly recyclable without any loss of quality. After iron and aluminium, copper is one of the most recycled metals. As much as 40 per cent of the copper used by European industries is produced by reusing old copper.

Recycling copper involves fewer steps than winning the metal from the raw rock. Moreover, to produce one metric tonne of copper from recycled metal uses just 20 per cent of the energy needed to mine, process and produce the same amount of copper from the original ore itself.

The environmental lesson for all of us is to take care when we dispose of our dead machines, defunct devices, out-of-date phones and crashed computers. Even though they are no longer working or wanted, there is much valuable stuff to be found in, and recovered from, our televisions and mobiles, laptops and gadgets. I was therefore going to end this brief history of copper with an environmental injunction: 'penny wise pound foolish' – given that old pennies, or 'coppers', were made of copper, albeit in the form of the alloy bronze. But I'm out of date.

Between 1797 and 1859, old one penny coins were indeed minted from pure copper. From 1860 until 1970, one penny pieces were cut from bronze, with a copper content of around 97 per cent. However, the price of copper, inflation and the move to decimalisation saw further changes in both the size and composition of the one penny piece. Since 1992, one penny coins have been lighter in weight and smaller in size. They are made of copper-plated steel, which makes them magnetic. In fact, only 6 per cent of the modern coin is copper, with the remaining 94 per cent being mild steel.

The history of the one penny coin runs in parallel with the rising price of copper. From pure copper to bronze to a thin-veneer of copper plating on a mild steel core, strictly speaking the one penny piece is no longer a 'copper'. Its days as a coin are probably numbered, perhaps signalling the end of 'bad pennies turning up' and 'spending a penny' when the need is urgent – although it's many a long year since a pee cost a penny. The coin's value lessens by the day. Many people see it as a fiddly nuisance when handling change. Contactless payments by plastic card and smartphones are becoming the transaction of choice for more and more people. Although the Age of Electricity and the Digital Age have seen copper itself grow both in price and importance, that very success has meant the gradual decline of copper in the 'copper'.

However, just to muddle things a little more, many higher-value coins do contain relatively high proportions of copper. For example, the £2 coin is made of about 75 per cent copper, with the remaining 25 per cent being composed of nickel and zinc, while the 'bi-metallic' £1 coin is made of nickel and brass with a high copper content. So, although there is regret and nostalgia about the likely demise of the one penny piece, copper lives on in our higher-value coinage, albeit in its alloyed form.

Chapter 5

Iron and Steel

Iron and steel conjure up images of strength and resolve. Political leaders love to metaphorically present themselves as women of iron, men of steel. Margaret Thatcher was the Iron Lady, immovable, resolute and determined. Sometimes with scorn, sometimes with favour, the Duke of Wellington's famous resolve earned him the title of the Iron Duke. Otto von Bismarck, the man who unified the small German states into the German Empire, became known as the Iron Chancellor.

From around 1912, the Georgian-born dictator Iosef Vissarionovich Dzhugashvili began to call himself Joseph Stalin. The Russian word for steel is *stal*. Joseph Stalin, Man of Steel. However, it was Winston Churchill in his 1946 speech who first saw an 'iron curtain' being drawn around the lands of the borders of the USSR, marking the beginning of what was to become the Cold War between West and East:

> *From Stettin in the Baltic to Trieste in the Adriatic an iron curtain has descended across the Continent... The Communist parties, which were very small in all these Eastern States of Europe, have been raised to pre-eminence and power far beyond their numbers and are seeking everywhere to obtain totalitarian control.*

Iron and steel are also used as adjectives, again to denote strength. We talk of those with a will of iron. Some people have nerves of steel. To have an iron grip is to enjoy total control. And those who plough on, focused and undeterred, show a steely determination, and make cast-iron promises.

We talk of an iron fist in a velvet glove when we want to describe someone who might appear gentle and soft but turns out to be hard and resolute, even brutal. The daunting triathlons known as the Ironman competitions involve a swim of 2.4 miles, followed by a bike ride of more than 100 miles, before leaping on to the road to run a full marathon. How's that for strength, stamina and resolve?

Metaphorically speaking, therefore, iron is one of our favourite metals. It also turns out to be the most abundant metal in the universe. Created by nuclear fusion deep in the heart of massive suns, iron is just one of the scores of elements blasted out into the cosmos as these huge stars finally die in the spectacular explosions that we see as supernovae. Subsequent recycling of the scattered stardust leads to the creation of new suns with their orbiting planets. The relatively heavy iron atoms generally accrete closest to their sun, and in the case of our own solar system, help form the inner rocky planets, including the Earth.

By total mass, our planet – third rock from the sun – is about 35 per cent iron. Most of this iron mass is concentrated in the Earth's core. The inner core is solid and made mostly of iron with some nickel. It has a radius of over 750 miles. This iron heart is surrounded by an outer core of a molten iron-nickel alloy with a thickness of 1,400 miles.

The vast currents that twist and swirl within this hot

iron-nickel zone help generate the Earth's magnetic field. This field deflects the sun's electrically charged solar winds. This in turn preserves the atmosphere's ozone layer, which filters out much of the sun's harmful ultraviolet radiation and allows life to flourish. The Earth's magnetic field also gives rise to the aurora borealis and aurora australis, the northern and southern lights, which appear when charged particles in the solar wind are attracted into, and interact with, the upper atmosphere above the poles.

Surrounding the core is the mantle. It accounts for 84 per cent of the Earth by volume. It is a region rich in a variety of rock types. When these various mantle minerals and rocks are analysed, they reveal that in terms of their composition, oxygen is the most abundant element, forming about 45 per cent by weight. The other three most common elements in this layer are magnesium (23 per cent), silicon (21 per cent) and iron (6 per cent). These three normally combine with the oxygen to form minerals known as silicates.

And shifting and shunting above the hot mantle's viscous currents are the tectonic plates of the Earth's crust, skin-thin in the grand scheme of things. However, it is on the rocky surface of this few miles of crust that we live and go about our puny business.

Analysis shows that about 5 per cent of the crust is composed of iron. It is found in many of the crust's minerals, including haematite and magnetite, garnet and iron pyrites, also known as 'fool's gold'. Although in absolute terms there are billions and billions of tons of iron present in one form or another in the Earth's crust, it is rarely found in its pure elemental state. It is generally locked away inside the complex chemistry of ferruginous minerals. Which is why, in spite of its abundance, it took so long for us to appreciate iron's presence, properties and potential.

However, there is an exotic exception to the general rule that iron is not found naturally in its raw metallic state. Each year, thousands of meteorites from outer space hit the Earth and its atmosphere. Many of the smaller ones burn up on entry and we see them as shooting stars. A few of the larger ones make it to the ground. Meteorites vary in composition. There are meteorites that are predominantly stony in nature. There are some that are a mixture of stone, iron and other metals. There are also those that are mainly made of iron with some nickel – the 'iron meteorites'. When they crash to Earth in areas of dry land and desert, they are more likely to be preserved and easier to spot, and of course in such places the iron is less likely to rust.

Although it wouldn't be until the Iron Age that men and women learned to smelt iron, a few ancient civilisations had found the metal in its raw state when they came across meteorites glinting on sandy desert pavements. They learned to fashion and shape this heaven-sent iron by heating and beating it.

In 1922, the archaeologist Howard Carter discovered the tomb of Tutankhamun. The preserved body was that of an Egyptian pharaoh who ruled between about 1332 and 1323 BCE. Three years into the archaeological search, two exquisitely crafted daggers were found among the burial wrappings of the young mummified king. They were made of iron, some nickel and a small amount of cobalt. X-ray analysis conducted by scientists in 2016 revealed that the iron-nickel alloy from which the daggers were made came from a meteorite. Ancient Egyptian beads, rolled, rounded and cylindrical, have also been found, again made from meteoric iron and nickel.

*　*　*

Some early pioneers were beginning to understand that pure iron could be smelted from rocks. The technology first appeared in the lands of Asia Minor, part of modern-day Turkey, Syria and Iraq, around 3,400 years ago. Over the following centuries, the ability to release iron by smelting rocks slowly spread west into Europe, although it did not reach Britain until about 2,800 years ago. These early breakthroughs marked the end of the Bronze Age and the beginnings of the Iron Age.

Equally important in the development of iron and steel production, of course, was where and how to find the ores that could be smelted to release the iron. Once it was realised that some rocks and stones when heated yielded iron, it was possible to search for rocks and stones of that type. Most iron-rich bearing rocks tend to be dark red, purple or steely-grey in colour. When they are exposed to the air and rain, the iron weathers and oxidises to a rusty-brown, giving yet another clue that iron might be present.

Certain geological conditions must be fulfilled before iron in rocks becomes sufficiently concentrated for it to be worthwhile mining. Most iron ores occur in sedimentary rocks – that is, rocks derived from the erosion and weathering of older rocks. As the weathering leaches out the iron from the minerals in the original host rocks, it eventually precipitates and settles out among the muds, silts and sands at the bottom of lakes or on the beds of the sea floor or beneath the waters of a boggy swamp. When these rusty iron-rich muds, silts and sands sink ever deeper beneath the layers above, they slowly compact and compress over the vast aeons of geological time to become sedimentary rocks suffused with iron-rich minerals.

When men and women of the Iron Age first discovered that metallic iron could be released by heating some soils and rocks in wood-fuelled fires, it was probably the more common, easily available iron-rich deposits that they exploited. Goethite (named after the German poet, Goethe, who loved to collect minerals), also known as 'bog' iron', is a form of iron hydroxide that forms when other iron-rich rocks are weathered. Goethite accumulates in soils and at the bottom of bogs.

Laterites are another iron-rich soil more often found in tropical areas where wet seasons are followed by hot, dry seasons. Here, lateritic soils often accumulate to significant depths. Laterites have a high iron oxide content and as a result they are typically coloured a rich red-brown.

Iron Age people soon developed increasingly effective ways of extracting the iron from these ores. Bloomeries were a key invention. A bloomery is a pit in the ground or a stack made out of clay or stone with a chimney on top. Air vents or pipes are inserted at the base. They allow strong draughts of air to be sucked in and higher temperatures to be achieved. Bellows also began to be introduced to increase the flow of air.

It was realised, too, that charcoal was a much more effective heat source than wood on its own. Produced by burning wood under low-oxygen conditions, charcoal is almost pure carbon. The bloomery is preheated by burning the charcoal. The furnace then begins its work, and the burning charcoal gives off carbon monoxide. This 'reduces' the iron oxides in the ore to metallic iron. The reaction involves the carbon monoxide removing the oxygen from the iron oxide to leave pure iron. Although the heat of the furnace isn't high enough to melt the iron ore, it is sufficiently hot to help release the iron from the rock itself, which is left behind as a 'slag'.

During this reduction process, the iron absorbs a little of the carbon from the charcoal and gas, but care must be taken that the proportion remains very low, otherwise the iron can't be worked. After the process has finished, a spongy mass of iron and some of the slag, known as a 'bloom', is left. The bloom is then reheated and beaten with a hammer to drive out the molten slag. Iron achieved in this way is described as 'wrought' (from the medieval Middle English for 'worked'), hence wrought iron.

Right up until medieval times, the bloomery and its charcoal fuel remained the way in which most iron was produced. As with all metals, iron gets stronger if it is hammered. People learned to hammer iron into ever more elaborate shapes for a widening range of uses. From beads and ornaments, axes and swords, sickles and scythes, wheel rims and ploughs, the skills of the blacksmith helped transform farming and fashion, travel and warfare. Indeed, in his 1858 lecture, the art critic John Ruskin recognised iron's 'power of bearing a pull, and receiving an edge … which enable it to pierce, to bind, and to smite' and so 'render it fit for the three great instruments, by which its political action may be simply typified; namely, the Plough, the Fetter, and the Sword'.

The growing importance and widening use of iron also led to the discovery of other minerals and rock types that were rich in the metal. The minerals magnetite and haematite are both forms of iron oxide. Siderite is an iron carbonate. These minerals can be found in a variety of sedimentary rock types, generically often referred to as ironstones.

* * *

The next major advance in iron-smelting technology came in 1709 when Abraham Darby replaced charcoal with coke as a fuel. Coke is made by heating coal in a low-oxygen environment. This produces a carbon-rich, porous material, ideal for burning and creating high temperatures. Darby first developed 'coke-coal' furnaces when he was working in – where else – Coalbrookdale, Shropshire. The low-impurity coal was mined locally. The result of using good-quality coke in his coke-fired blast furnace was to produce iron of high quality.

For many, this innovation marks the beginning of the Industrial Revolution. In 1777, Abraham Darby's grandson, known as Abraham Darby III, began construction of the world's first cast-iron bridge over the River Severn. This is the famous Iron Bridge, less than a mile from the original Coalbrookdale ironworks. It opened on 1 January 1781.

Darby's coke-fired innovation soon became the basis of all iron and steel production. It originally meant that the ideal sites for building blast furnaces were places where both iron ore and coal (for coke) were available and could be mined nearby. Indeed, until recently it was always the case that iron-making took place in areas where there were iron ores. For example, iron had been smelted in and around the Weald in south-east England since before Roman times. Plentiful supplies of iron-rich sands and clays as well as wood from trees to make charcoal ensured a one-time thriving industry right through until the end of the eighteenth century and the beginnings of the Industrial Revolution.

In Britain, by the early nineteenth century, iron and steel production was shifting to the north. The Jurassic iron-stones of Cleveland in North Yorkshire and coal from the nearby Durham pits saw the growth, beginning in 1851, of major iron and steelworks along the River Tees from Middlesbrough to Redcar.

Steel for both the Sydney Harbour Bridge (opened in 1932) and the smaller version that became the Newcastle–Gateshead bridge over the River Tyne (opened in 1928) was produced by the famous Dorman Long Steelworks in Middlesbrough. Both bridges were designed by the London firm of Mott, Hay and Anderson. Although the Tyne bridge was first to open, the Sydney Harbour Bridge was designed first – but being much larger, it took longer to build. Then, in 2015, after 164 proud years, the very last blast furnace at the Redcar works was shut down, marking the end of major iron and steel production in the region.

* * *

Today, most of the ores mined to make iron and steel come from rocks that have been given the name *banded iron formations*, sometimes knowns as BIFs. Geologically speaking, these rocks are fascinating. They are rich in iron minerals and feel unexpectedly heavy when you pick up broken chunks. They have a reddish, rusty, stripy appearance when exposed to the elements, can be found on all continents and are very, very old. And most interesting of all, they tell us something about what kind of place our world was more than two billion years ago; they reveal secrets about the oceans, life and the Earth's atmosphere in those ancient days. They also play a key part in the story of life, a story that eventually allowed oxygen-breathing life forms, including us, to evolve.

The BIFs are sedimentary rocks, although given their great age, they are often lightly 'metamorphosed', that is, they have been subject to great heat and pressure during those times in which they have lain deep beneath the Earth's surface. Originally, they were deposited at the bottom of the sea. The bands are the result of fine alternations between layers of darker iron oxide minerals and lighter layers, usually of fine sand and silica in the form known as chert.

The ages of these rocks vary depending on where in the world they are found, but most are between 3.0 and 1.8 billion years old. This far-distant time was known as the Pre-Cambrian. The land then was a barren place without life. Rivers would pour off the continents, carrying the weathered remains of volcanoes and mountains out into the surrounding seas. Sands and silts settled on the ocean floors. The river waters were also saturated with many dissolved minerals and elements, including those of iron, which were added to the salty chemical cocktail of these early oceans.

There was, of course, life in the sea, most of it simple and single-celled. It was around 2.5 million years ago that some of these oceanic bacteria and algae evolved to perform the extraordinary trick of turning sunlight into chemical energy to help fuel their metabolism. This process is known as photosynthesis. Light energy is used to convert water, carbon dioxide and minerals into living organic compounds, while at the same time releasing free oxygen into the surrounding environment.

Up until that point in the Earth's history, there was very little oxygen in the atmosphere. It was the appearance of these early photosynthesising cyanobacteria that gave rise to what we now call the 'Great Oxygenation Event'. This began around 2.4 billion years ago. It fundamentally changed the chemistry, biology and ultimately the geology of the planet. And it made fire possible.

Iron is an element that is chemically very reactive in the presence of oxygen, where, depending on how much oxygen is available, it forms a variety of iron oxide minerals. These include magnetite, haematite and siderite. The huge backlog of dissolved iron in these once oxygen-free Pre-Cambrian oceans began to react with this newly produced, bacterially generated oxygen gas to create vast amounts of iron oxide of one kind or another.

The bacterial blooms waxed and waned so that there were periods of higher levels followed by lower levels of oceanic oxygen, resulting in parallel periods of more, then less, iron oxide precipitation. During the slacker phases, the only sediments accumulating were the silts, sands and silica precipitates. These processes of iron oxide precipitation followed by silica sedimentation continued for hundreds of millions of years, resulting in the formation of hundreds of feet of thin, alternating bands of dark iron oxides and pale cherty silicas stretching over thousands of square miles of the seabed, creating the banded iron formations, the BIFs.

At first, so much dissolved iron had accumulated in the oceans that nearly all of the initial oxygen produced by the algae and bacteria was reacting with the dissolved iron to form the heavy iron minerals that were slowly sinking to the bottom of the oceans to form the BIFs. This had a fascinating consequence.

Because nearly all of the bacterially produced oxygen was initially ending up as various iron oxide precipitates, there was little left to bubble up and out into the Earth's early atmosphere. For the first couple or more billion years of the planet's 4.6 billion-year history, there was little free oxygen in the air. The atmosphere was mainly one of nitrogen, carbon dioxide, methane and a few other volcanically erupted gases. No atmospheric oxygen meant no oxygen-breathing life on the land, whether plant or animal.

And then, around 1.8 billion years ago, when most of the iron in the oceans had been oxidised by the oxygen produced by the photosynthesising bacteria and algae, the oxygen they were still generating was now free to escape into the atmosphere. Atmospheric oxygen levels slowly began to rise. The time of the banded iron formations was mostly over. The only dissolved iron now entering the seas was that which was still being brought down by the rivers, but there was nothing like enough to react with all of the oxygen being generated by the oceanic algae and bacteria.

The photosynthetically generated oxygen was now free to saturate the seas and circulate in the atmosphere. With more and more free oxygen in the waters and the air, oxygen-metabolising life forms began to evolve across the planet. The death of the banded iron formations saw the birth of creatures that could breathe oxygen.

However, it would be another billion years or more before atmospheric oxygen reached present-day levels. The next major evolutionary event occurred around 800 million years ago, when oxygen-breathing life forms began to evolve at a faster and faster rate. And then, 540 million years ago, evolution took a massive leap. This has been called the 'Cambrian Explosion'. Myriads of new life forms – plant and animal, on land and in the sea – burst onto the scene.

Meanwhile, the BIFs were lying buried, heated and compressed under thousands of feet of younger deposits. There they continued to lie, subject to the usual pushes and pulls, rises and falls, buckles and bends of the Earth's geological forces. It is only when these processes bring the BIFs to the surface that the rocks become subject to erosion and weathering. And, of course, they also become ripe for mining, particularly open-cast mining on vast scales using heavy machinery whose size and power are frighteningly awesome.

* * *

Worldwide, well over 2,000 million metric tonnes of iron ores are mined annually, most of this from the BIFs. The outcrops of the BIFs can stretch for hundreds of miles. They vary in their iron-rich content, but huge open-pit mines can make even the lower-grade ores economic. The world's greatest producer of iron ores is Australia. Most of the 800 million metric tonnes it currently produces annually is loaded onto trains, sent by rail to ports and shipped to China. However, some of the world's biggest individual mines are in Brazil, which comes second in the league table of major producers, followed by, at the time of writing, China, India, Russia, South Africa, Ukraine, Canada and the United States. Most of these are open-cast mining operations.

The world's biggest underground iron ore mine is at Kiruna in northern Sweden. A very different set of geological processes gave rise to these Scandinavian deposits. The Kiruna ores were formed around 1,800 million years ago following intense magmatic and volcanic activity. Iron-rich solutions precipitated iron oxides in the form of a very pure magnetite-apatite mix, containing more than 60 per cent iron. After Ukraine, Sweden is Europe's biggest producer of iron ores, and the ores are of a very high grade.

* * *

Before they can be turned into iron and steel, mined ores are first crushed and then sorted. High-grade ores containing magnetite need less refining. Lower-grade ores need to be refined to remove various impurities and contaminants.

These various refining methods are called 'beneficiation'. They involve crushing, washing with water to float sand and clay away, magnetic separation, pelletising and sintering. At the end of these processes, the refined ore is ready to be sent off to blast furnaces.

A blast furnace is a tower-shaped structure, made of steel, and lined with refractory, or heat-resistant, bricks. The furnace is loaded from the top with the iron ore, limestone (which acts as a flux) and coke. At the bottom of the furnace, very hot air, often enriched with oxygen, is blasted into the mix through nozzles called tuyères. The coke burns in the presence of the hot air. The oxygen in the air reacts with the carbon in the coke to form carbon monoxide. The carbon monoxide then reacts with the oxygen in the iron oxide ore to form carbon dioxide. This 'reduction' process leaves much purer iron behind. At this stage, the carbon content in the iron is still relatively high at around 4 to 5 per cent.

The molten iron sinks to the bottom of the furnace. The limestone flux combines with the remaining rock and its impurities to form a slag. This is lighter than the iron and floats on the surface of the liquid metal. The iron is drawn off at the bottom while the slag is tapped and removed at points higher up the furnace. As molten iron and slag are removed, more blended ore is introduced at the top in a continuous process. Blast furnaces typically run non-stop for years. The only time they are allowed to cool is when the refractory bricks that line the furnace finally begin to crumble and need to be replaced. Or when the iron and steelworks are closed, as happened at Redcar.

The melted iron is either cast into ingots called 'pigs' or sent off to be mixed with other metals or materials to form one or more of the many alloys of iron. Cast iron is produced by the slow cooling of remelted pig iron in the presence of coke and limestone. One of the distinguishing features of cast iron is that it has a relatively high carbon content of between 1.5 and 5 per cent. Cast iron is not very malleable: it is brittle and hard; it cracks easily; it has low resistance to tension, but is very good under compression; and it has no plasticity (and therefore cannot be forged or rolled), but is easily machined. Depending on the carbon content, its hardness and low malleability make it ideal for making railings, machine frames, pipes, bearing housings, cylinders for steam and vehicle engines, and flywheels.

Wrought iron is the purest form of iron. With less than 0.2 per cent carbon, it can be up to 99.8 per cent pure. It is prepared from pig iron by burning out carbon, silicon and other impurities. Although it is not suitable for casting, because it is very ductile and malleable, it can be drawn into wires and pipes, or rolled out into sheets.

The most common alloy of iron is steel. Steel is a malleable alloy of iron and between 0.2 and 2 per cent carbon. As an alloy of iron, in terms of its carbon content, steel lies midway between wrought iron and cast iron. Steel is most commonly made by the 'basic oxygen process' (BOP). Pure oxygen is blown into a bath of molten blast furnace iron to which more scrap can be added. The oxygen reacts with any impurities present, including carbon, silicon, phosphorus and manganese. The oxidised impurities escape as hot gases, although the process is controlled to ensure that just the right amount of carbon is left to produce the type of steel wanted. Electric arc furnaces can also be used to purify blast-furnace iron to produce steel.

Other metals and materials can also be added to create a wide variety of speciality steels. The range of uses for iron, and particularly steel, is staggering. So much of the modern world is based on, or built around, steel, from the smallest needle to the biggest bridge. Steel girders hold up skyscrapers. Steel can be honed to make razors with edges exquisitely thin and so sharp that they slice through paper and hair. Steel can be pressed and shaped to produce the body of a car or the door of a fridge. It can be drawn and hammered into long lines of railway track. It can be magnetised or stretched to the thinness of a wire. Machine parts, drills, saws, ploughs, train wheels, chisels, ball bearings – the list goes on and on. But we might just stop and say a little more about one special kind of steel: stainless steel. Stainless steel resists corrosion and can be found fashioned into kitchen sinks and jewellery, cutlery and saucepans, razors and vats, surgical instruments and washing-machine drums.

Harry Brearley was a metallurgist based in Sheffield in the years just before the First World War. He was born in Sheffield in 1871, and his father was a steelworker. Steel was in Harry's blood, iron was in his soul. One of his jobs as a young researcher was to see whether the hardness of the steel used to make rifle barrels could be improved by adding various elements of one kind or another to the iron alloy. It was all trial-and-error stuff. One day he noticed that one of his abandoned efforts had not rusted like so many of his other trial pieces. It had retained its silvery metallic sheen. Harry was curious. This particular steel alloy was one in which he'd added chromium as well as a little carbon. Brearley realised he might be on to a winner. With more experimentation, he eventually made what we know today as stainless steel.

It took a while to get the proportions right, including the addition of nickel, but every time you use a shiny knife and fork, or spoon your pudding off a plate without suffering the tangy taste of the cutlery itself, it's stainless steel you have to thank. In 1914, Brearley moved to Thomas Firth and Sons, and so began the Sheffield stainless steel industry, still famed today for its cutlery.

Inevitably, given the strong minds, vested interests and entrepreneurial talent that stainless steel was beginning to attract, there were early squabbles, disputes and tensions. In part, these were resolved when Brearley and Firth agreed to set up the Firth-Brearley Stainless Steel Syndicate. But in spite of these initial difficulties, credit was eventually given where credit was due. In 1920, the Council of the Iron and Steel Institute presented Brearley with the Bessemer Gold Medal awarded for outstanding services to the steel industry. Quite why the medal was made of gold rather than steel, I don't know.

* * *

Today we use twenty times more iron and steel than all other metals put together. In 2019, China produced more than 900 metric million tons a year, almost half of the world's annual production of nearly 1,900 million metric tonnes. Then some considerable way behind come India, Japan, the USA, Russia and South Korea. Germany remains western Europe's biggest producer, pouring out just under 40 million metric tonnes in 2019, while the once great iron and steelworks of the UK could manage only 7 million metric tonnes in that same year.

But there are costs. Vast tracts of land are stripped and scarred as huge diggers and trucks remove millions of tons of rock and ore, year on year. In Brazil, rainforests are bulldozed flat. Soils are removed. Local flora and fauna are lost. Rivers become polluted and fish die. Indigenous people lose their homes and way of life. Coal is needed to make coke, in the process of which huge amounts of carbon dioxide, a greenhouse gas, are produced, adding to global warming. And although the slag created during the blast furnace stage does have several uses, including as a road metal and filler in concrete, much of it ends up in landfill.

And as for iron and steel themselves. They don't last forever. They rust. They crumble. They stain.

* * *

As iron decays into one or more of its various oxides, we discover it has a second life as a metaphor. Rusty when you're out of practice. Old and rusty when things no longer work. America's 'Rust Belt' describes the great sweep of country from New York, across Pennsylvania, West Virginia, Ohio, Indiana, the Lower Peninsula of Michigan, Illinois and Wisconsin where so many industries, including iron and steelmaking, went into steep decline after 1980. Rusting away as machines lie abandoned and unloved. Rusting away as people give up on life. Marie de Hennezel's book in which she meditates on ageing has the wonderful title *The Warmth of the Heart Prevents Your Body from Rusting*.

However, not everything about rust needs to be seen as negative. Indeed, there is a deep, life-affirming quality to rust that takes us all the way back to the beginnings of complex, oxygen-breathing life forms during the Great Oxygenation Event that took place over two billion years ago.

The red of our red blood cells results from the presence of a complex iron protein known as haemoglobin. When red blood cells reach the lungs, the iron in the haemoglobin binds with the oxygen that has been breathed in by the animal to form oxyhaemoglobin. The iron in the blood has become oxygenated. Rusty, if you will. The oxygen-rich blood is then pumped around the body. When the oxyhaemoglobin meets a cell saturated with carbon dioxide produced as the cell metabolises, a complex chemical exchange occurs in which oxygen is delivered to the cell to fuel its metabolism. At the same time, the cell's accumulated carbon dioxide is transferred to the haemoglobin to form deoxyhaemoglobin. When the deoxygenated red blood cells return to the lungs, the exchange process is reversed again. Carbon dioxide is released from the cell into the lungs and exhaled. We breathe in oxygen. We breathe out carbon dioxide. That's life.

The nineteenth-century polymath and champion of all things arts and craft John Ruskin wrote a famous essay in praise of rust. His thoughts were originally delivered as a lecture in the spa town of Tunbridge Wells in February

1858. He titled his talk 'The Work of Iron in Nature, Art and Policy' and began his lecture with these opening lines:

You all probably know that the ochreous stain, which, perhaps, is often thought to spoil the basin of your spring, is iron in a state of rust: and when you see rusty iron in other places you generally think, not only that it spoils the places it stains, but that it is spoiled itself – that rusty iron is spoiled iron... [and that] we suppose it to be a great defect in iron that it is subject to rust.

Ruskin then tells his audience they would be wrong to think this. He proceeds to list the virtues of rust. He confirms that 'the breath of life' is possible only because of blood's clever trick of first rusting then unrusting as it does its work around the body. 'Is it not strange,' he wonders, 'to find this stern and strong metal mingled so delicately in our human life, that we cannot even blush without its help?'

And so he goes on, reminding us that the rocks from which we extract so much bounty, and the soils in which we grow our crops and graze our animals, are rich in and enriched by rust. The many oxides of iron and their complex mingling with other elements give us the soft yellows, subtle reds and rich browns of the newly ploughed field. The cosy warmth of red-brick buildings and the bright vermillion of pantile roofs that please the eye owe their appeal to rust. Then, when iron's work is done, rust returns it to the earth from whence it came. Nature's due is paid and balance is restored.

However, in spite of Ruskin's romance with rust, 150 years later he would have been horrified at the environmental cost that iron and steel production continues to exact. At heart he was a conservationist. Today, most iron ore mining operations are vast. Deep pits are dug as monster machines cut and scrape their way down into the terraces of the iron-rich rocks. The red dusts created by these voracious diggers cover and colour the land for miles around. In 2016, 7.2 per cent of the world's annual greenhouse gas emissions, those drivers of global warming, were pumped out by the iron and steel industry.

Nevertheless, our bustling world of tower blocks and bridges, cars and trains, machines and tools is inconceivable without iron and steel. Our appetite for the metals sees no end in sight. They are two more metals that lies at the heart of modern life. Iron and steel production make up 95 per cent of global metal production. There is 'iron in the soul' of our civilisation. And so, concludes *The Economist* magazine, the development of a process that turns raw earth into iron and steel really should merit as 'a high spot on a list of mankind's most ingenious achievements'. Without iron and its alloys, so much of the modern world would literally collapse.

Chapter 6

Concrete

Concrete all too often gets a bad press. Things made out of it are seen as big and brutal. It doesn't always weather very prettily. But it's everywhere. Office blocks, car parks, sewage pipes, bridges and roads can all be constructed out of concrete. It is ideal for building monolithic structures such as dams, sea defences and nuclear shields. It has helped build the infrastructure of the modern world, allowing millions to live safe, clean and efficient lives. In so many cases, it literally forms the foundations of twenty-first-century life. And because it is made and laid while fluid and wet, it is an extraordinarily versatile material.

Builders and civil engineers recognise many types of concrete. Any material that is made up of an *aggregate* bound and held together by a paste or *cement* might be defined as a concrete. Aggregates range from coarse sands to gravels and pebbly stones. The cement might be made of bitumen (for concrete used to make roads) or calcium aluminate (where the concrete has to be strong and resistant to heat and corrosion). However, the most common and ubiquitous cement is lime-based cement, especially Portland cement. It is this kind of cement upon which we shall concentrate.

Although earlier Middle Eastern civilisations certainly used cement-based products in many of their buildings, it is the Romans who are generally credited with being the first to make and use concrete on a large scale.

The fascinating properties of lime had been known for centuries. Lime is produced by heating rocks made of calcium carbonate ($CaCO_3$) such as limestone (hence the name). The fierce heating process drives off carbon dioxide from the calcium carbonate rock to leave behind a white powder of calcium oxide (CaO), also known as quicklime. If water is then added to quicklime, it undergoes a chemical reaction, giving off heat in the process. First the calcium oxide is 'hydrated' to produce a paste of 'slaked lime'. Slaked lime then reacts with carbon dioxide in the air. This brings it back to calcium carbonate, which eventually sets hard and white. Neat. Although more complex and elaborate in practice, it is this basic chemical process that underlies much of lime, cement and concrete production.

The Romans discovered that by adding volcanic ash and pumice to the lime and water mix, they could produce not only a concrete that could be moulded and shaped but one that would also set under water. As their confidence in using concrete grew, Roman architects and builders became more and more adventurous in their designs. Arches and domes began to span ever-wider spaces. The Pantheon in Rome, for example, was built nearly two thousand years ago. Its huge, unreinforced concrete dome, supported on columns of granite, remains one of the world's wonders and an architectural marvel.

* * *

Joseph Aspdin was born in Leeds in 1778. He was a brick-layer, and he obviously knew and thought a lot about his materials because in 1824 he submitted a patent entitled 'An Improvement in the Mode of Producing an Artificial Stone'. He called his invention 'Portland cement' as he thought that his solidified concrete had the appearance of the famous Portland stone, an oolitic limestone quarried from the Jurassic rocks found on the Isle of Portland in Dorset, out of which, for example, St Paul's Cathedral and the National Gallery are built.

Aspdin took rocks quarried from his local Carboniferous Limestone beds. These rocks help form the Pennines. When they outcrop on the surface, they create the distinctive land-scape seen in the wonderful limestone scenery to be found in places such as Malham Tarn in the Yorkshire Dales and the Derbyshire Peak District. Having quarried the lime-stone, the next stage was to crush it into a powder.

The powder was then heated. This turned the calcium carbonate into calcium oxide in a process known as 'calci-nation'. Aspdin next added clay and water to the mixture before drying it out. The solid remains were broken into lumps that were placed in a furnace and again subjected to great heat, which drove out the carbon element as carbon dioxide gas. The remaining 'calcined' product, or 'clinker', was finally ground and rolled into a fine powder – Portland cement. The cement was now ready to be mixed with aggregates and water to make concrete. Because of the high costs of transporting limestone rocks, most cement-producing factories continue to be built on or near the quarries themselves.

Joseph Aspdin's son, William, originally worked with his father in Leeds, but they fell out. William, generally seen as a bit of rogue and a liar, moved to Rotherhithe in south-east London and began producing his own cement using chalk, which is even richer in calcium carbonate than Carboniferous Limestone.

Chalk rocks can be found all over and under Kent, offering a ready, easily accessible source of raw material. William burnt his chalk 'limestone' and clay mix at a much higher temperature than the method pioneered by his father. This resulted in a cement that contained more calcium oxide and was a much more versatile product. Its greater hardness and strength meant that it was suitable for making high-quality concrete. As it set much more slowly than his father's product, it also gave builders more time to fashion and form the concrete into whatever designs the architects had envisioned. William Aspdin's method produced what we now refer to as 'modern' Portland cement.

Over the years, both the composition and manufacture of Portland cement have undergone many refinements. Today, manufacturers carefully control the chemical mixtures of calcium, silicon, aluminium, iron and other elements to produce a series of complex minerals of calcium and aluminium carbonates and silicates. The basic ingredients to make the cement include limestone, chalk, marls and even shells. These calcium carbonate-based rocks are mixed with clay, shale, slate, blast furnace slag, silica sand and iron ore. The whole concoction is then heated to very high temperatures in a cement kiln, and the resulting clinker is ground into a fine, light-grey powder composed of calcium oxide, calcium silicates, aluminium oxide and small amounts of other oxides. This is what we now think of as powdered cement.

Concrete is formed when water is added to Portland cement and an aggregate of, say, sand or gravel. Typically, a mix is about 10 to 15 per cent cement, 60 to 75 per cent aggregate and 15 to 20 per cent water. The wet concrete can be poured into 'forms' for moulding to create walls, supports, floors, pillars, chimneys and so on.

For slabs, concrete is left to stand until the surface moisture film disappears. Wooden or metal 'handfloats' are used to smooth off the concrete. 'Floating' produces a relatively flat, but slightly rough texture. If a smooth, hard, shiny surface is required, floating is followed by steel trowelling. The water in the mix reacts with the cement, which begins to harden and bind with the aggregate. Once it has hardened, the concrete forms a solid in the shape formed by the mould.

Technically speaking, then, once poured, concrete doesn't actually dry out, it sets. The cement reacts with the water, triggering a series of chemical reactions that lead to the formation of solid concrete. It is therefore critical when mixing concrete to get the amount of water exactly right. Too much or too little and the concrete is weakened, liable to crumble and, when put under stress, collapse. On the morning of 14 August 2018 in the Italian city of Genoa, a 200-metre section of the busy Morandi concrete road bridge collapsed. Part of the bridge was built above residential housing. As hundreds of tons of concrete and cars tumbled earthwards, buildings were crushed. Forty-three people were killed. Six hundred more were left homeless.

The building industry has an insatiable appetite for concrete. China is currently the world's biggest producer and consumer of concrete. Since 2003, it has produced more cement every three years than the USA managed throughout the twentieth century. In 2019, worldwide, over 4 billion metric tonnes of cement were produced, of which China manufactured more than half. When mixed with aggregates, this produces more than 10 billion metric tonnes of concrete every year, equivalent to more than 1 metric tonne for every man, woman and child living on the planet. This makes concrete the single most widely used man-made material in the world.

* * *

Limestone and chalk are sedimentary rocks. Sedimentary rocks are ultimately derived from the erosion and weathering of pre-existing rocks, including all three of the major rock types – igneous, metamorphic and previously formed sedimentary rocks. The eroded material to make sedimentary rocks might settle along a river valley, be blown along as desert sands, sink to the bottom of a lake or lagoon, or be swept out to sea, where it settles in sinking ocean-floor basins, layer upon layer, over millions of years.

Some eroded minerals dissolve in water. These weathered chemicals might be taken up by the plants and animals that live in the rivers, lakes and seas. When the plants and animals die, their remains also sink and accumulate as organic matter or biological detritus. They, too, can form sedimentary rocks.

Mineral-rich waters can also warm up under fierce suns on hot days. The brines evaporate, leaving behind their salty precipitates to form yet another type of sedimentary rock – the evaporites. The Triassic salt rock beds beneath the Cheshire Plain offer a fine example. Another is the dolomitic Magnesian Limestone of the Permian period found in north-east England, formed of coral reefs and by the evaporation of the Zechstein Sea that was washing over north-west Europe more than 250 million years ago. The minerals extracted from these deposits contributed to the growth and international success of Teesside's chemical industries.

Not surprisingly, sedimentary rocks are most commonly found as layered beds across the surface of the Earth's crust. However, although sedimentary rocks cover more than 70 per cent of the planet's continental surface, they make up less than 10 per cent of the total volume of the crust itself, if we also include oceanic crust. The rest of the crust, often tens of miles thick, is made up of igneous and metamorphic rocks.

Limestones, including chalk, are mainly the accumulated remains of creatures whose structure and shells were made of calcium carbonate or calcite. Limestones make up about 10 per cent of all sedimentary rocks. Sandstones, siltstones, shales and clays make up most of the rest.

Carbonate sedimentary rocks offer a window onto not only the planet's geological history but also the evolution of life on Earth. There are many ages and types of limestone, but we shall look at just three, which have been and continue to be used in the manufacture of cement and concrete: Carboniferous Limestones, Jurassic Limestones and chalk.

Some 358 million years ago, much of what is now North America and northern Europe, including the British Isles, was part of a continent known as Laurasia. It straddled the tropical latitudes just *south* of the equator. Warm, shallow seas, lagoons and tidal flats washed over much of this area, which up to that point had been a land of hot, dry deserts. The waters were rich in carbonate minerals. The seas also teemed with life. Many of the plants and animals had shells and structures made of calcite. Coral reefs flourished.

Over the next few million years, the sinking sea floors became the resting place of the dead remains of these shelly creatures and coralline crags. As the shells and calcareous muds slowly accumulated and compacted, they became cemented by more calcite as it precipitated out of the calcium carbonate-saturated waters. These limey deposits first began to be laid down around 355 million years ago, before finally giving way 323 million years ago to a very different landscape of vast river deltas of coarse sands. These delta sediments would eventually lead to the formation of the slightly younger Millstone Grits. But it was beneath these grits that the earlier calcite-rich marine sediments were beginning their gradual mutation into the strong, compact and dense rock we know today as Carboniferous Limestone.

This limestone is hard and resistant to weathering. Often seen embedded in its smooth, pale grey/white weathered surfaces are the lacy remains of ancient corals, the shells of brachiopods and the stems of crinoidal sea lilies. It is Carboniferous Limestone that creates the areas of robust, often wild upland scenery found along the spine of the Pennines, in the southern Lake District, over parts of North and South Wales, the Mendips – including Cheddar Gorge – and south-west England, and across the central plains of Ireland.

The rock's toughness has led to its use as a building stone for mansions, monuments, houses and halls. It can be seen wandering for mile after mile as drystone walls that criss and cross hillsides. It is quarried for roadstone and railway ballast. It is used as a flux in the production of iron and steel. And, of course, it is a source rock for making cement, which is used in the manufacture of concrete.

Throughout these Carboniferous times and beyond, the Earth's tectonic plates continued to shift and shunt over the planet's surface. By around 200 million years ago, northern Europe and North America, including the British Isles, now found themselves about 30 to 40 degrees latitude *north* of the equator. This was the beginning of the Jurassic period, a time when Britain was once again a place of coastal lands and warm shallow seas. For a while, muds and clays were the main sediments to settle on the floor of these early Jurassic waters. Entombed in these fine-grained rocks were the remains of many types of sea creatures, including ammonites, molluscs and the occasional marine reptile – such as those dug out of Dorset's crumbling cliffs by the redoubtable Mary Anning during the first half of the nineteenth century.

However, by the Middle Jurassic, 174 to 164 million years ago, in many parts of what is now south and east England, thick layers of carbonate deposits accumulated between the clays, silts and sands as they settled at the bottom of the subtropical seas. Under compaction, 'lithification' and aeons of time, these became the Inferior (meaning below and therefore older) and Superior (meaning above and therefore younger) Oolitic Limestones. These rocks are shelly and 'oolitic'.

Oolites (from the classical Greek for 'egg' and 'stone') are very small, rounded grains of calcium carbonate that form on the sea floor when the waters are saturated with calcium carbonate. The calcitic oolites form around tiny grains of sand or broken bits of shell that act as a 'seed'. As the seeds wash to and fro in the warm tidal seas, layer upon concentric layer of calcium carbonate begin to accrete around the grains to form rounded oolites. Similar processes can be seen today in the seas around the Bahamas.

Jurassic oolitic limestones crop up on the Isle of Portland, where they are known as Portland stone, in honour of which Portland cement was named. The limestones then run as a strip from Dorset heading north-east, where they rise to form the Cotswold Hills, up through the East Midlands and on into Lincolnshire.

For hundreds of years, the warm, creamy-coloured stone has been used to build both humble cottages and grand houses. In the city of Bath, the local oolitic limestones are more honey-coloured. Their fine grain and uniform structure have made the stone a popular choice for architects, even to this day. Bath's world-famous Royal Crescent shows Bath stone at its finest. Once again, like all carbonate rocks, these Jurassic Limestones can be pummelled, mixed with clay, heated and ground to make cement that goes into the manufacture of concrete.

The Jurassic period came to an end 145 million years ago. The ancient continental blocks of North America, Europe and Asia continued to drift north, dragged by the mighty forces of plate tectonics.

By the beginnings of the Late Cretaceous, just under 100 million years ago, much of northern Europe and beyond was hovering at around 40 degrees north of the equator, at the same latitude as modern-day Spain, southern Italy, the Mediterranean and the Azores. Global sea levels were 200 or 300 metres higher than they are today. This area of northern Europe was covered by seas that were growing wider and deeper. They were now free of the muds, silts and sands washed down by the faraway Cretaceous rivers. All that rained down on the sea floor was a slow, steady drizzle of the dead skeletons of tiny planktonic algae known as foraminiferans and the even smaller coccoliths that bloomed in the sunny waters hundreds of metres above.

The tiny plates and remains of these small creatures were composed mainly of the white mineral calcium carbonate. And so, millennia after millennia, for the next 30 million years, the seabed continued to sink beneath layer upon thin layer of the soft white calcareous ooze. In these warm, balmy waters, the rate of seabed deposition was very, very slow. Every 1,000 years the limey ooze thickened by just one more centimetre.

With further downward movements of the Earth's crust, the oozes compacted and hardened. They turned into rock, the pure, soft, very white limestone that we know today as chalk. Chalk rock occurs in vast undulating sheets that rise and fall across north-western and central Europe, from the

Crimea, across the Paris Basin and fields of Belgium, and on to the white cliffs of Dover and the English Downs, over to Salisbury Plain, up to East Anglia, and on to Yorkshire and Flamborough Head, whose white cliffs jut out so magnificently into the North Sea.

The first person to suggest a name for these distinctive white rocks was the extravagantly named Belgian geologist Jean Baptiste-Julien D'Omalius d'Halloy, the only son of an ancient and noble family. In 1822, when describing and mapping the gently rolling landscape of the Paris Basin, he talked of the terrain as being crétacé – chalky land. The Latin for chalk is *creta*, and hence the period in which these chalk rocks were deposited has become known as the Cretaceous. It was to the chalklands of Kent that William Aspdin took himself after he'd fallen out with his father, and it was chalk that he used as he developed and refined his father's original cement-making recipe.

* * *

The properties of Portland cements made them ideal for mixing concrete. It is concrete's remarkable ability to remain plastic and workable when newly mixed, but strong and durable when hardened, that has helped make it the world's most used and versatile building material. Throughout the late nineteenth and twentieth centuries, architects began to design more and more buildings and structures made out of concrete.

It was the introduction of steel bars into the setting concrete that further reinforced and increased its strength.

French manufacturers laid claim to the invention of this 'reinforced' concrete, but it was German engineers who became experts in understanding its physical and mechanical properties, allowing them to calculate what loads and forces an architectural design could support. Reinforced concrete permitted new designs to become possible. Floors in ever-higher skyscrapers, beams to support walls, and posts to hold up roofs all began to appear.

Concrete's ability to be 'formed' into any shape meant that architects could dream up designs that might otherwise have been impossible to realise. The French architect Auguste Perret's apartment building in Paris was completed in 1904. It became famous for being one of the first buildings not to hide, but rather to deliberately celebrate, its concrete structure. He is often credited as being one of the twentieth century's pioneers in designing and building in concrete. Reinforced concrete seemed to be pointing the way to a new kind of architecture. Another French architect, Le Corbusier, for example, began to create buildings of extraordinary originality and beauty based on the now ubiquitous 'post and slab' system of concrete construction.

Modernism's celebration and uncompromising use of 'raw' concrete led to it being labelled 'brutalist architecture' after the French 'brut', meaning raw. Here, there was no attempt to disguise concrete's qualities and looks. Surfaces were typically left rough and unfinished, leaving the pattern of the wooden boards between which the concrete had been poured. Buildings had a massive, heavy, blocky look to them. No longer need a building be a complex assembly of different materials and parts of stone, steel, brick and wood. It could be all of a piece, all one continuous structure, sometimes referred to as *monolithism*, where, writes the architectural historian Adrian Forty, 'walls are just up-ended floors'.

Begun in 1959, it took fourteen years to build the Sydney Opera House, with its enormous pre-cast concrete shells. It was designed by the Danish architect Jørn Utzon, but tensions between him and some of the commissioners and politicians eventually led to the Dane's resignation in 1966. In spite of the growing acrimony during its construction, Utzon's work, along with the structural know-how of the engineers of the Ove Arup team, is now regarded as not only innovative and stunningly beautiful, but ahead of its time.

But despite concrete's many, many aesthetic successes, it continues to provoke mixed feelings. To talk about concrete is to talk about modernity and the ambivalence that goes with it. Even as early as 1925, the artist and architect Henri-Marcel Magne was able to write that so much of the 'raiment' of modern architecture is reinforced concrete.

Concrete's aesthetic weakness as well as its strength is that it is, as the American architect Frank Lloyd Wright said, a 'mongrel' material, neither stone nor plaster, neither brick nor tile, neither iron nor steel. Adrian Forty goes on to add:

> *Efforts to 'naturalize' concrete are endless – to give it the smoothness of polished marble, the density of limestone, the laminar quality of wood, every one of these ways of finishing concrete are attempts at representing the otherwise formless and inarticulate results of the process as 'natural'. What all of them confirm is that concrete permanently awaits deliverance from the category of the 'unnatural'.*

It was the nineteenth-century polymath John Ruskin who felt that buildings and their form should remain true to the innate character of the materials out of which they were built. This became known as 'structural rationalism'. The design and form of a building should be based on the nature of the things out of which it was being constructed. Brick should appear as brick, stone as stone. Materials shouldn't dissemble.

But what is the natural form of concrete? What is it that architects are supposed to remain true to when they build with concrete? Answers to these questions slowly began to be found as architects, and just as importantly engineers, understood that concrete's ability to take on any shape or form was its true character and indeed its virtue. In concrete, anything became possible.

However, when design and appearance go astray, concrete becomes a synonym for all that is wrong with modernity and urban life. The appearance of many natural materials such as stone and wood improve with age. They mellow and mature. But concrete all too easily weathers into ugliness. Stained, mouldy-green, crumbling, discoloured office blocks and multi-storey car parks are attacked as 'concrete monstrosities'. Acre after acre of high-rise, 'post and slab' blocks of flats, graffitied walkways, and dark and dank underpasses create a forbidding, drab-grey, dystopian cityscape, a 'concrete jungle'. Concrete is everywhere, and everywhere it looks the same. It was only a short political jump to see concrete as the architecture of poverty, where uniformity and soullessness drained people of their humanity.

After the death of Joseph Stalin in 1953, his immediate successor was Georgy Malenkov, but he was replaced as leader six months later by Nikita Khrushchev. As soon as he was in post, Khrushchev acknowledged the USSR's serious housing problem. He made a speech in 1954 announcing that wherever possible constructions should be made out of concrete. It was cheaper than metal, quicker to assemble than brick, and more versatile than stone. He had in mind a concrete socialist utopia.

But as Adrian Forty points out, the policy led to some extreme absurdities. For example, in the Baltic states, where there are vast timber forests, even the telegraph poles were built of concrete. They marched in disciplined lines across the country and through the redundant timber of the endless pines. In the cities, huge numbers of standardised, pre-cast concrete apartment blocks began to appear. The urban landscape became grim and grey. Nor were Western cities immune to this widescale use of pre-cast concrete and panel-based building, especially for social housing. Concrete was the universal material of choice during the boom years that erupted after the Second World War. However, it remained largely unloved, especially by those who had to live in the high-rises and hurry home along the fetid walkways.

Today, architects and engineers have many types of concrete with which to work. Each type varies in terms of its appearance and properties. In the hands of the best exponents, it has resulted in some stunning constructions. It seems invidious to pick out any in particular, but here are just three, arbitrarily chosen, to remind ourselves that concrete can be exceptionally beautiful as well as unforgivably ugly.

The ethereal Millau Viaduct in southern France with its concrete piers and supporting cables is breathtakingly elegant. It was designed by the British architect Lord Norman Foster and his team, and the French structural engineer Michel Virlogeux.

Perhaps one of the most sensuous builders in concrete, steel and glass was the Iraqi-born architect Zaha Hadid. Her twenty-first-century designs bend and flow, curve and caress, zig and zag. The Heydar Aliyev Cultural Center in Baku, Azerbaijan, completed in 2012, sees her seductive vision at its most fluid. She died, aged sixty-five, only four years after its opening.

Completed in 1986, the Lotus Temple in New Delhi, India, has a flower-like 'lotus' structure. The lotus flower is often regarded as a unifying symbol in India's religions. However, the geometry of the building was so complex it took the Iranian-American architect Fariborz Sahba and his team more than two and a half years to complete the drawings alone. The concrete frame and pre-cast concrete-ribbed roof incorporate the white, marble-clad 'petals' that when viewed arouse feelings of cosmic transcendence. The petals are arranged in clusters of three to form nine sides, surrounded by nine pools of water. Light pours in as the central petals of the 'lotus' peel open to let in the sun and sky. Little wonder that it has become one of India's most visited and celebrated buildings.

★ ★ ★

However, with concrete being the most widely used man-made material, it is not surprising to learn that cement manufacture and concrete production exact a heavy cost on the well-being of the planet. Huge quarries and pits have to be dug to extract the billions of tons of limestone, sand and gravel needed every year. They cut into hillsides and scar the landscape. Only when they have been abandoned can they be used for landfill. Or, if simply left, time often proves to be a gentle healer. In many cases, nature has slowly crept her way back in, softening the ravaged rocks, carpeting the quarry floors with flowers and grasses. Some of the oldest quarries have become nature reserves where birds, butterflies, bees and other bugs have found sanctuary.

It is in the manufacture of cement itself that we see the greatest environmental impact. The very high temperatures required to turn limestone and clay into cement are normally generated by burning carbon-based fuels such as coke, oil or natural gas. This process produces huge amounts of carbon dioxide, a greenhouse gas. In order to produce one ton of cement, approximately one ton of carbon dioxide is generated. This, along with the decarbonisation of the limestone itself, contributes roughly 8 per cent of the world's annual man-made carbon dioxide emissions. There is no getting around the fact that the process of making cement releases carbon dioxide in huge volumes. However, there are attempts to substitute 'greener' fuels and technologies for the cokes, oils and gases currently used to heat the limestone mix. So far, though, only minor gains have been made to cut down on the greenhouse gases that cement manufacturing creates.

Sand is concrete's other main ingredient. Huge amounts are needed to keep up with the demands of the construction industry. The world consumes around 50 billion tons of sand and gravel aggregates a year. On the positive side, there is a lot of sand in the world. It can be found on the seabed, deposited along river valleys, dumped by ice to form glacial moraines, and blown into dunes by desert winds.

On the downside, many of the sands that are removed are those whose loss is environmentally damaging. The loss of marine and river sands by dredging can destroy aquatic habitats. Sand extraction can lower the water table and pollute supplies of drinking water. In some countries, there is a secretive, illegal trade in beach sands. Criminals plunder the coastal sands of poor countries and export them to sand-hungry rich countries. The removal of beach sands causes untold environmental harm. Coastal erosion increases. Wildlife habitats are destroyed. Local fishing is upset.

There is also the problem of what to do with concrete once it has weathered, is no longer fit-for-purpose and unwanted. Although concrete is strong and durable, it does deteriorate over time. Alkali–silica reactions and carbonation cause it to crumble and crack from within. The waste from demolished concrete buildings often ends up in landfill sites. Some is ground up to be used as an aggregate or pounded to provide road metal. The reality, though, is that concrete is not easily reused or recycled. As a result, it falls towards the bottom of most league tables of sustainable materials.

There are one or two glimmers of hope. The idea of a self-healing concrete is being developed to increase its lifespan. The technique involves embedding self-activating limestone-producing bacteria into the building material. When water seeps into the concrete, it activates the microorganisms, which secrete limestone into the hairline cracks. Other innovations are exploring the use of different aggregates, such as recycled plastics and blast-furnace slag, in attempts to reduce concrete's heavy carbon footprint.

* * *

We'll end this discussion of concrete on an even darker note by considering what can be done with it when novelists and film-makers get their imaginations working. Some have a fondness for writing about mobsters who seal the fate of their enemies by entombing them in concrete and cement. People who have upset the bad guys are liable to end up buried in concrete under newly constructed bridges. Engineers will tell you that dead bodies in concrete supports do little to improve the strength of their bridges. Further 'accidents' are likely to occur in the years down the line.

Euphemistically, other victims are given 'concrete shoes' to wear before being thrown into the sea, where they can 'sleep with the fishes'. Death by encasing the feet of those who betray the 'mob' was used by E. L. Doctorow in his 1989 novel, *Billy Bathgate*, in which a 15-year-old Bronx boy bears witness to the murder of a man thrown into the East River wearing 'cement shoes'. Even more extreme is to wear a 'concrete overcoat'. In his famous novel *Slaughterhouse-Five*, Kurt Vonnegut mentions a magazine story's mistaken belief that the character Montana Wildhack was dead. The article said that 'she was wearing a cement overcoat under thirty fathoms of saltwater in San Pedro Bay'. Her diving expedition was not thought to have been voluntary.

Chapter 7

Glass

Glass is a solid. But you can see through it. Isn't that strange? It's so easy to take glass for granted that we often forget that it's a material with very curious properties.

Not all glasses, though, are transparent. For example, there's 'petrified lightning'. This is a natural glass that forms when lightning bolts strike the ground, including sandy ground. The tubes of vitrified, opaque rock that result are formally known as *fulgurites*, from the Latin *fulgur*, meaning lightning or thunderbolt. Although they can sometimes reach tens of centimetres in diameter and can be even longer in length, they are typically hollow.

Silica-rich volcanic rocks that erupt on the surface and cool so quickly that they don't have time to crystallise can also become glassy. Obsidian is perhaps the best example. It is black in colour and, like flint, another type of silica; it is very brittle and easily fractures to give razor-sharp splinters that can be used, as Stone Age craftsmen and women discovered, for making scraping knives, arrowheads and lethal spears.

Glasses fashioned by nature reveal the basic conditions under which a glass is likely to form. You need to melt or liquify the right minerals and cool them very quickly so there is no time for crystals to grow. In essence, glasses retain the characteristics of a liquid, albeit one that has perversely become solid. Or more accurately, a liquid that has become so viscous on cooling that to all intents and purposes, it has become solid.

In general, as they cool, liquids lose energy. They become more sluggish. Their readiness to flow slowly decreases. In most cases, when a cooling liquid reaches a critical temperature, it freezes, becoming solid, forming a tightly interlocking structure made of crystals. This is what happens when water turns to ice.

However, no such change occurs when glass in cooled. The liquid glass simply gets stiffer and stiffer until it becomes as stiff as a solid. Only under special cooling circumstances will a glass 'freeze' and form crystals. In this case, the glass is said to have devitrified and it can no longer be regarded as a glass. So, in rather simplistic terms, we might follow the science writer Terence Maloney and define glass as 'a rigid liquid'.

It is because glass doesn't form a crystalline structure that it retains the disorderly, random molecular structure of a liquid. There is no regularity or pattern in the distribution of the molecules that make up glass. And like many liquids, this lack of structure – this absence of regular crystalline surfaces off which light might reflect – allows light to make its way through glass relatively unimpeded. This, coupled with glass's failure to absorb energy in the visible light spectrum, means that we can see right through it.

A number of minerals lend themselves to forming a glass, but we'll stick with simple silica – that is, silicon dioxide (SiO_2) – as it is far and away the most common raw material. We generally know it better as sand. Many of the most common sands are composed of trillions of small grains of silica, generally seen in its crystalline form and known as quartz.

* * *

Sand is fragmented rock where the grains are smaller than gravel but larger than silt. Weathering and erosion slowly break down all rocks and their constituent minerals into smaller and smaller bits. Quartz is a particularly hard mineral. It is also difficult to break down chemically. This partly accounts for its prevalence at the end of the erosion process.

Not all rocks contain quartz. Volcanic rocks such as basalt contain next to no silica or quartz. Granites, by contrast, and by definition, are made up of between 10 and 50 per cent quartz. The other major mineral present in granites is feldspar. Feldspars make up between 50 and 90 per cent of any granite. Indeed, they are one of the most abundant rock-forming minerals in the Earth's crust. They are the big pink-to-whitish-grey crystals that give granites their distinctive look and make them perfect for kitchen worktops, tombstones and building veneers.

As we learned in Chapter 3, granites are most typically found in the hard, rugged hearts of great mountain chains, both old and new. They are generated in the region where two tectonic plates converge. Continental crust gets pushed up in the crash zone to form long mountain chains. Deep beneath these rising mountains, one of the advancing tectonic plates gets driven beneath the crust of the less dense continental plate. It is 'subducted', and it is in these subduction zones and mountain roots that molten granitic magmas begin to form.

Huge plumes of molten rocks, especially those rich in silica, begin to rise beneath the fractured and fissured crustal rocks. There these molten, silica-rich rocks sit, slowly cooling over millions of years. Very slow cooling allows very big crystals to grow. This accounts for the large, beautiful feldspar and quartz crystals we see in granites. And because the huge masses of cooling granite, known as 'batholiths', were once molten and generated deep underground, they are referred to as plutonic igneous rocks, named after Pluto, the Roman god of the underworld, and *ignis*, the Latin for fire and heat.

In the British Isles, huge granite batholiths underlie the Lake District, the Cheviots, the Cairngorms of Scotland, Devon and Cornwall, South Wales and the north Pennines. They represent the deep-rooted remains of ancient mountain chains, long since worn away. Millions of years of lands rising and rocks eroding now see these granitic mountain roots being exposed at the surface to give us the rugged scenery of many of Britain's uplands.

Even as mountain chains are being pushed up, the wind and the rain, the ice and the snow, begin to wear them down. In 1830, Sir Charles Lyell published one of the first major books on geology, *Principles of Geology*, in which he described the restless, ever-changing surface of the planet. The poet Alfred, Lord Tennyson had read the book and used its geological imagery in his 1850 poem, *In Memoriam*, to capture the inexorability of change in all things:

The hills are shadows, and they flow
From form to form, and nothing stands
They melt like mist, the solid lands,
Like clouds they shape themselves and go.

Given time, then, all solid land melts like the morning mist. Gradually, the relentless rasp of the elements reveals every mountain's granite heart. Granites, though, are tough. They are among the most resistant rocks to the processes of weathering. But bit by bit, over the millennia, their fractured grains fall into mountain glaciers, hurtle down in tumbling torrents, rumble along the bottom of swelling streams, get dumped across flat river plains, become blown into desert dunes, fan out over river deltas, settle on sinking seabeds, and fringe continental coastlines with sandy beaches. And as quartz is one of the more resilient of these weathered granitic minerals, it is quartz grains that form the bulk of these sandy sediments.

Geologically speaking, it is also worth reminding ourselves that many of the sands that settle on sinking seabeds, pile up along river valleys or get blown into desert dunes will themselves be subject to tectonic forces and geological processes. Those that sink and become covered by more sediments might find themselves buried deep beneath the crustal surface. Under heat and pressure and over time they will be turned from loose sediments back into compacted rock. These sands will become sedimentary sandstones. They too might be subject to erosion and re-release their grains of quartz to create a new generation of loose sands.

* * *

Quite when the knack of turning silica sands into glass was first discovered isn't entirely clear. As with iron and copper, ancient Middle Eastern civilisations certainly began to get the idea that molten sands could produce glassy-like materials. Egyptian and Mesopotamian beads made of glass have been dated to around 4,000 years ago. However, it was not really understood how glass could be heated, shaped and turned into objects of one kind or another. It would be another couple of thousand years before the beginnings of a deliberate glass technology. By the time of Ptolemaic Egypt around 2,300 years ago, it was becoming possible to make glass rods. These would be reheated and shaped into dishes and bowls.

A hundred or so years later, again in Egypt and the Middle East, craftsmen and women worked out that if you blew down an iron tube – a blowing iron – into a blob of hot, viscous glass and rotated it as you did so, you could make glass 'balloons' and turn them into bowls, crude bottles and bulbs. While hot, the glass could also be moulded, bent, twisted or rolled into the required shape.

Another key development was the discovery that by pressing a hot, blown-glass bulb flat against a cold surface, it was possible to produce small plates of glass. Unfortunately, the plates were thick, with a rough, cloudy finish. Nevertheless, here was the beginning of the idea of flattish glasses that could be fitted to glaze over window spaces. It would be some time before this technique became sufficiently refined for glazed windows to become commonplace.

The Romans also discovered that the addition of small amounts of different metallic oxides could produce glasses of different colours. Copper oxides would result in red or green glasses, cobalt oxide gave a blueish tinge, iron oxides led to glasses of green and brown, while antimony oxides stained glass yellow.

By the Middle Ages, not only was stained-glass technology improving, the manufacture of clearer, more transparent glasses was also getting better. However, it was still not possible to make plates of window glass sufficiently big to cover large areas. The only way to glass over a big window space was to use smaller panes and link them together in a framework made of lead dividers.

In the hands of the most skilled glaziers, huge leaded windows were built in churches and cathedrals. By using a whole colour palette of stained glasses, patterns and pictures not only adorned the building, they also allowed sunlight to pour into the naves, aisles and transepts. Glorious examples of stained-glass windows began to spring up all over medieval Europe. The sunlight streaming through the leaded lights of the cathedrals of Chartres and Canterbury and the minster at York still have the power to take the breath away with their luminescence and beauty.

Elsewhere in Europe, around 1200 CE, Venice became a major centre for the manufacture of glass. The Venetians had learned some of the techniques that had recently been developed in Syria and Egypt and refined them even more. By the 1400s, the best Venetian glassmakers were making glass of great clarity, glass that was vibrantly coloured, glass that was gilded and even enamelled. Although the Venetian craftsmen and craftswomen tried to keep their skills secret, their technological innovations soon spread throughout Europe. Glassware for drinking as well as for ornamentation became established across the continent.

* * *

As glassmakers continued to learn more about glass and its properties, their innovations began to play a key role in the development of the natural sciences. For a long time, it was recognised that rounded blobs of glass, or indeed certain transparent crystals, could magnify any object viewed through them. By the late thirteenth century, glassmakers had learned that by grinding and polishing pieces of glass they could produce curved surfaces that had a lentil or 'lens' shape. When objects were looked at through these lens-shaped glasses, they appeared larger. This marked the beginning of the optical industry and the invention of the first purposefully manufactured spectacles, or 'glasses', to help with close work and reading.

Further experimentation eventually led to the invention of the microscope and the telescope. The scale of the universe was suddenly stretched in both directions. A whole new world was discovered at the level of the very small, a world teeming with life that was restless and full of the most exquisite complexity and structure.

In the other direction, telescopes took the human eye out millions and billions of miles up and away to the far reaches of the heavens. The place of men and women in the scheme of things took on a whole new perspective. Glass was beginning not only to open our eyes to possibilities unimagined, it was also causing us to stop and think about who we might be in the vastness of the universe. The philosophical and religious ramifications of these realisations continue to reverberate to the present day.

There is uncertainty about who first invented the microscope and the telescope. However, the technological hub for these early optical instruments was undoubtedly the Netherlands. Dutch spectacle-makers were already skilled in grinding high-quality lenses, and by the 1590s, the first microscopes had begun to make their appearance. In the compound microscope there are at least two lenses. The objective lens is positioned close to the object and produces an image that is picked up and then magnified further by the second lens, called the eyepiece. Improvements quickly followed. Galileo Galilei, for example, made his own *occhiolino*, or 'little eye', in 1609.

The English scientist Robert Hooke used his microscope to examine the structure of snowflakes, fleas, lice and plants. When looking at plants and other living things, he noticed the presence of tiny biological, self-enclosed structures. He called them 'cells' from the word *cella*, the Latin for 'small room', such as those lived in by monks. His detailed observations appeared in a beautifully illustrated book, *Micrographia*, published in 1665.

The rapturous reception of such works meant that science never looked back as the understanding of optics and the power of microscopes continued to grow. Over the next four centuries, lenses began to play a key role, not just in the advance of science, but also in the development of cameras for photography, and by the twentieth century for making moving pictures for the cinema and television.

A similar story occurred with the invention of the telescope. Again, there is debate about who made the first telescope, but once more we have to look to the skills of Dutch lens-makers for some of the earliest examples of these 'spyglasses'. Certainly, by the early 1600s, the Netherlands was recognised as an optical centre of excellence. The first person to apply for a patent for a telescope was a Dutch eyeglass-maker named Hans Lippershey in 1608, although there was fierce rivalry between Lippershey and some of his fellow countrymen, including Zacharias Jansen and Jacob Metius, over who made the best if not the first telescope, or indeed microscope.

And again, Galileo Galilei makes his way into the optical story. He had heard about the 'Dutch perspective glasses' and soon set about making his own telescope. By 1609 his improvements allowed him to magnify distant objects by up to twenty times. What Galileo did, which was revolutionary, was turn his telescope skywards to look up to the heavens. This marked the birth of astronomy as an empirical science. In no time at all he was seeing craters on the moon, rings around Saturn, moons spinning about Jupiter, and a band of light arcing across the clear black, night sky that we now know is an end-on view of our own galaxy, the Milky Way.

As a scientist, it became clear to Galileo that Copernicus was right. The Earth could no longer be seen as the centre of the universe. Along with the other planets, it revolved around the sun. For the Catholic Church, this was heresy and went against the teachings of the Bible. Under threats from the Church, Galileo had to back-pedal, although he was rumoured to have whispered that whatever the priests thought about the Earth remaining as the static centre of the world, *sotto voce* astronomers would say, 'And yet it moves'.

Over the centuries, other great scientists followed in Galileo's starry wake. Johannes Kepler, Isaac Newton, William Herschel and his sister Caroline, and Edwin Hubble looked at the night skies and began to understand that the universe was not just big, but, as Douglas Adams wrote in *The Hitchhiker's Guide to the Galaxy*, 'vastly, hugely, mind-bogglingly big'. It gradually became apparent that it is not just our galaxy that is full of billions of stars, but that there are billions of galaxies each made up of billions of stars. The science of cosmology began to take on shape and scale.

Physics, too, got a boost when George Ravenscroft, a seventeenth-century English glassmaker, discovered that the addition of lead oxide produced a more sparkly glass in which colours could be dispersed. This was not only the beginning of the lead crystal and cut-glass industry, it also led to improvements in the properties of other optical devices, including the prism.

Lead oxide glasses have a higher refractive index than conventional glasses. This means that they are suitable for making prisms, and prisms are good for splitting light into its various wavelengths, which is to say all the colours of the rainbow. In his investigations into the properties of light, using prisms and lenses, Newton, much to the dismay of the poet John Keats, 'unweaved the rainbow'. So began the scientific journey that would lead to our modern understanding of light as an electromagnetic wave particle, with a speed limit – the speed of light – that finds its way into so many areas of basic physics and the very nature of matter itself.

Perhaps less glamorous, but equally important in science's advance, was the development of the humble glass container. Just as lens-makers were getting better at grinding glass, glassmakers were learning to shape better phials, beakers and in time test tubes and other laboratory equipment. Glass vessels allowed early chemists to see what was going on as their compounds and liquors fizzed and foamed. More crucial still, glass proved to be relatively chemically unreactive. Things could boil and bubble away within walls of glass without involving the glass itself. Right up to the present day, glass vessels have proved an invaluable, even if taken-for-granted, tool in the growth of so many sciences and medical practices.

* * *

The common-or-garden window, of course, is one of the first things we think of when we consider glass and its many uses. It's easy to forget what a marvel of technology it is as sunlight pours into our homes and brightens our lives. Glass windows let the light in and keep the wind out. It wasn't always so. Throughout medieval and Tudor Britain, most windows were of stone or timber construction. They had unglazed openings that might be covered with oiled cloth, paper, wooden shutters or even thin sheets of horn – anything to keep out the cold and stop the drafts. Glazed windows were for the rich and, even then, they used only small panes of glass set in a lead-strip latticework.

Until the sixteenth and seventeenth centuries, the most common method for making flat sheets for window glass depended on the skills of the glass-blower. A large blob of glass was twirled onto the end of a blowpipe. Air was then forced down the pipe and into the hot glass blob until it ballooned to form a spherical globe roughly 50 centimetres in diameter. This glass globe was then reheated and attached to a long iron rod known as a pontil. The hot, viscous blob of glass was squashed against a flat surface made of marble, steel or brass. The flattened glass was then reheated, lifted and twirled on the end of the pontil in order to rotate it. The spinning, centrifugal force caused the viscous glass to flare out into a thinning, hot disk, and the spinning had to be maintained until the cooling glass was sufficiently rigid to retain its flattened, circular shape. However, the round plate was rarely entirely flat – the glass was thicker at the centre and thinner at the edges. The plate's surface was therefore rippled with a series of concentric rings, usually with a blob or crown at the centre, hence it is often described as 'crown glass'. It was possible to spin sheets up to a metre or more in diameter, which were then cut to size. You can still see this type of glass in the ripply, rounded glass windows of old houses and olde worlde sweet shops.

It was crown glass that allowed the introduction of sash windows. However, because this type of glass was still expensive to make, the most popular type of window remained casement with leaded glazing. By the nineteenth century, new methods of window glass production began to be developed. This meant that the panes could be made larger, clearer and cheaper. Molten glass could be poured onto cool, flat surfaces or into a mould where it could spread out into thin slabs. Or hot glass could be rolled by machine into relatively flat, thin sheets that could be polished to make plate glass for windows and mirrors.

Built in 1851 as part of the Great Exhibition, the Crystal Palace was assembled from cast-iron frames and 300,000 panes of cylinder-rolled flat glass. The Exhibition, indeed the glass palace itself, was a celebration of Victorian engineering and manufacturing ingenuity. It was the first major building in the world to use glass as the main component.

However, as far as window panes are concerned, *float glass* has become the modern method of choice for most purposes. The process for making float glass was pioneered by Alastair Pilkington at his St Helen's factory in Lancashire. After several years of experimentation, float glass was first manufactured commercially in 1959.

Let's look at the journey sand has to take from being extracted by diggers to ending up as a large pane of float glass.

Most of the sand used in industry is produced by either scooping up deposits of loosely consolidated sands or ripping up and then crushing weakly cemented sandstone rocks. The bulk of this sand is used for aggregates in the building industry.

Silica sands for glassmaking generally have to be of a higher quality than building sand. Some of the best sands in the British Isles for making glass are found around Chelford in Cheshire. These are bright, white glacial sands of high purity and consistent grain size. They were deposited towards the end of the last Ice Age about 20,000 years ago.

Once collected, the sand producers have to prepare the sand so that it reaches a standard acceptable to the glass manufacturers. First, the sand is washed. Grains that are either too small or too big are removed. Froth flotation, gravity separation and magnetic separation help take away any heavy minerals, especially iron, that might be present. Chemical washing might also be necessary before the sand can finally be taken off to the factory to make glass.

Silica sand on its own doesn't melt until it reaches a temperature above 1,700°C. The chemical bonds that hold silicon and oxygen atoms together in a crystal of quartz are very strong. A lot of heat energy is needed to break these bonds and turn the solid silica into liquid glass. It is possible to lower the softening point, and hence lower the viscosity, through the addition of other minerals, including sodium oxide and sodium carbonate. This produces the most common type of glass – soda-lime glass. In the float glass process, these raw ingredients, which can include recycled glass, are continuously fed into a furnace and heated to around 1,500°C until the glass softens and begins to flow like honey or treacle.

As the molten glass leaves the furnace in a continuous strip, it is gradually cooled to a temperature of 1,100°C before finally flowing out onto a bath of molten tin, which is at a temperature of 232°C. *Floating* on the liquid tin, the molten glass spreads to form a flat, level sheet surface. Any bubbles of air present pop to the surface to leave a perfect, unsullied sheet of glass. The glass is then allowed to cool slowly. The whole process depends on a precise balance between hot glass, hot metal and a hot atmosphere. Depending on the size of the molten bath of tin, sheets of glass can be made in almost any size. Once the float glass has cooled to about 100°C, it is rolled along the production line and cut to the required size. Float glass is free of distortion and can be manufactured to thicknesses that vary from centimetres to only a few millimetres.

The creation of float glass led to an architectural revolution. The ability to make very large panes of perfect glass meant that a building could be a tower of glass as well as a skyscraper of steel and concrete. There are many excellent, often spectacular examples of glass-clad buildings, including the Shard in London, the Dancing House in Prague and, perhaps most unexpectedly, the extraordinary, the slightly mad Basque Health Department Headquarters in Bilbao, Spain.

* * *

The twentieth-century science of glass has given rise to many remarkable new materials. Toughened glasses and laminated glasses that don't shatter can be fitted as car windows. Glass fibres that have helped speed up the digital world and allow us to take light into the darkest corners of the human body as doctors carry out the most delicate of keyhole procedures. Glass fibres for building boats and insulating houses. Photochromic glasses to protect the eye. And because it has a very high electrical resistance, glass as an electrical insulator.

The glass industry has also developed glass ceramics. These arise from the carefully controlled process of devitrification of special borosilicate glasses, better known to most of us under the commercial trade name of 'Pyrex'. When photosensitive glass is heated above its usual temperature of formation, it turns opaque and becomes very strong and resistant to thermal shocks. It can withstand extremes of cold and heat. You can fill a Pyrex dish with raw ingredients and cook them in a hot oven. None of which is to forget the more prosaic but equally important everyday glasses out of which we drink water and enjoy wine, the jars in which we store our pickles and jams, the bottles in which we buy our beer and spirits, the mirrors in which we look to comb our hair or adjust our clothing, the light bulbs and tubes that brighten our homes.

Glass, unlike most other manufactured materials, is usually fashioned into the finished article by the glass manufacturers themselves. We have seen this with the production of float glass. But it's also true of bottles, jars and dishes. Glass, therefore, lends itself to continuous processes of production, and hence speed, efficiency and cost saving. In bottle-making factories, for example, the production line begins with sand, soda and recycled glass (known as 'cullet') being fed into a furnace at one end and coming out as glass bottles at the other.

Huge amounts of energy are expended scooping up the sand, transporting it to factories, melting it and turning it into glass. The manufacturing side of the industry therefore has a fairly heavy carbon footprint, contributing to the world's annual emissions of greenhouse gases and global warming.

However, environmentally speaking, glass does have several redeeming features. Glass containers can be washed and reused again and again – and even if they aren't reused, in theory, once made, glass is infinitely recyclable. Glass's relative lack of chemical reactivity and inertness also means that once produced it doesn't pollute the environment with toxic chemicals. From an environmental point of view, the substitution of plastic-based bottles, jars and containers for glass ones has been very bad news.

In short, along with double-, even triple-glazed windows, reusable and recyclable bottles, and its inert chemical character, glass is one of the material world's better, if not totally good, green guys.

* * *

Glass also makes for good metaphors, similes, idioms, jewellery, works of art and book titles. Its key characteristics of transparency, brittleness, refractivity and reflectivity help us think about life and the human experience, what can be seen but not reached, what is pure but also fragile, what can splinter and break, what can refract and distort, and what can mislead and deceive.

You certainly shouldn't throw stones if you live in a glass house. But how else are women to break that glass ceiling that allows them to see the opportunities, rewards and promotions up there but in practice prevents them passing through that transparent but gender-discriminating barrier? Too much politics speak and management talk is smoke and mirrors stuff, designed to obfuscate and mislead – an *Alice*

Through the Looking Glass world. We need to see clearly, not through a glass darkly.

And what of those fragile folk whose hearts of glass have been cruelly shattered by an unfaithful lover? But maybe you are better off without your faithless friend. See your glass as half full, not half empty. This option was rarely taken by Charles VI of France, whose mental health was certainly fragile. He was prone to delusions in which he thought he was made of glass and might easily shatter if he physically came into contact with other people. In the event, he remained in one piece until his death in 1422, aged fifty-three.

Chapter 8

Aluminium

Although iron is the most abundant metal in the universe, coming in at 0.11 per cent, aluminium is the most common metal found in the Earth's crust. Oxygen (46.6 per cent) and silicon (27.7 per cent) make up almost three-quarters of the crust's elemental composition, followed by aluminium (8.1 per cent) and iron (6.3 per cent).

Most of this crustal aluminium is locked up in complex minerals known as aluminium silicates. These include the shimmering, iridescent micas common in many metamorphic rocks, and the large, beautiful feldspar crystals that we see in granites and which shimmer brightly on a sunny day. At the other end of the size spectrum are the tiny clay minerals. These are complex aluminium silicate hydroxides that make up muds and shales, soils and clays. We have already met kaolinite and mullite in Chapter 3 when we were thinking about pottery and bricks and what they are made from.

However, unlike iron, which has been worked as a metal for almost three thousand years, aluminium, in spite of it being the third most abundant element in the crust, wasn't discovered and isolated until 1825, when the Danish physicist Hans Christian Ørsted announced the results of his exploratory experiments. There were many, though, who doubted whether he had in fact managed to produce a new element of pure aluminium. In 1827, the German chemist Friedrich Wöhler visited Ørsted and it was agreed between the two men that Wöhler should continue refining the Dane's original experiments. By 1845, Wöhler was able to produce enough of the metal to begin investigating its properties. This marked the beginning of aluminium's rise as one of the industrial world's key metals.

Aluminium gets its name from its various hydrated sulphate salts. These are known as *alum*. Naturally occurring alums have been found and employed for the past couple of thousand years. They have been and continue to be used in medicines and as mordants to help fabrics take on a dye.

Even before Ørsted's and Wöhler's successes, many scientists, including Antoine Lavoisier in France and Humphry Davy in England, had suspected alum was some kind of salt of an unknown metal. But in spite of all their efforts, they could not isolate the element in its pure form. Nevertheless, anticipating its eventual discovery, it was Davy, a Cornishman, who was first to give the putative metal the name 'aluminium' as he attempted to extract it from the alums with which he was experimenting.

It took the rest of the nineteenth century to learn how to produce aluminium in sufficient quantities to bring the price down. The next challenge was to think about what it might be used for, other than jewellery and *objets d'art*.

The eventual industrial success of aluminium lies in its chemical and physical properties. It is silvery-grey in colour and a third as light as iron, though in its pure form not as hard. Whereas iron melts at 1,538°C, aluminium turns liquid at 660°C. Although not as good as conducting heat and electricity as copper, it is better than iron. But taken together, aluminium's particular characteristics make it an extremely useful and versatile metal.

As its various properties were recognised, the number of uses to which aluminium was being put increased dramatically towards the end of the nineteenth and beginning of the twentieth century. Pots, pans and plates began to be manufactured from aluminium. Various alloys were created, making the metal stronger without gaining too much extra weight. It began to appear in ships and the fabric of the first cars and planes. The demand for aluminium was on the increase.

★ ★ ★

As befits its name, the very first methods of producing aluminium involved doing various chemical and physical things with alum, which was itself obtained by treating aluminium mineral-rich rocks with sulphuric acid. But if the production rate was to be scaled up, new and bigger sources of the raw ingredients would be needed. Various ores were tried, but it was soon agreed that *bauxite*, which was both plentiful and accessible, was best suited for most large-scale economic mining operations.

One of the first places in which the ore was found was around the village of Les Baux-de-Provence in south-eastern France. In 1858, aluminium was successfully extracted from the ore. So, in honour of the place where the rock was first transmuted into metal, the ore was given the name 'bauxite'. And 150 years or more later, bauxite remains the world's primary source of aluminium.

Today, most of the world's bauxite is mined in regions that lie in wide belts that circle, or once circled, the equator. It has to be remembered that the further back in geological time you go, the more the continents have shifted thanks to tectonic plate activity, with lands that used to lie in the equatorial tropics having been carried this way and that across the planet. This means that many present-day deposits of bauxite might not find themselves anywhere near the equator.

In the tropics, heavy seasonal rains leach the more soluble rock minerals of calcium, sodium, magnesium and potassium. Aluminium minerals are less soluble. During dry periods, the more salty, soluble metal solutions are drawn upwards as surface moisture evaporates. They creep up through the rock, where they eventually collect on the surface. When the next rainy spell occurs, they are washed away into rivers and out to sea. This process of hot-wet followed by hot-dry followed by hot-wet leaves behind the *less* soluble rock minerals of iron, aluminium and silicon. Thousands and thousands of years of annual hot and dry seasons produce a gradual surface build-up of iron and aluminium-rich silicates, oxides and hydroxides. These are known as laterites.

The exact composition of a laterite depends on the type of underlying rock upon which the weathering and leaching processes took place. Granites, for example, are richer in aluminium minerals than those of iron, so laterites that form above these feldspar-bejewelled rocks have higher concentrations of hydrated aluminium minerals, including gibbsite ($Al_2O_3.3H_2O$ or $_2Al(OH)_3$), diaspore ($AlO(OH)$) and boehmite ($Al_2O_3.H_2O$). It is these laterites, rich in aluminium minerals, that are known as bauxite. Although some very old Pre-Cambrian bauxites do occur, most commercial bauxite deposits are found in the Cretaceous, Palaeogene and Neogene periods, with ages ranging between 145 million and a mere 2 million years.

Again, bauxite rocks vary in exact composition. However, by definition all have high levels of aluminium-rich minerals, particularly gibbsite. Australia leads the world in production, followed by China, Guinea and Brazil (figures from 2019). Because bauxite forms on or close to the surface, it is usually strip-mined. Each year, worldwide, 40 to 50 square kilometres of land and forest are ripped and stripped in pursuit of the metal, all of which has a huge environmental impact. Bauxite layers are typically between 4 and 6 metres thick. The rock is broken up by blasting, drilling, ripping and digging by monster machines. The rubbly rock is then transported to the refineries, where it is first crushed and then washed to help rid it of unwanted minerals, including clay.

* * *

Huge amounts of energy are needed to produce pure aluminium from bauxite ore. Although all metal-processing industries consume enormous quantities of energy, aluminium is particularly demanding. In most cases, aluminium's manufacture involves a two-stage process, a fact many of us first learned about when we were being taught chemistry at school: the Bayer process to produce alumina (aluminium oxide) followed by the Hall–Héroult process to produce pure aluminium.

The Bayer process involves heating the washed and cleaned bauxite in a solution of caustic soda (sodium hydroxide, NaOH). This takes place in a pressure vessel at a temperature of 150 to 200°C. The heating produces a super-saturated solution of sodium aluminate referred to as the 'pregnant liquor'. After a series of filtering, separation, cooling and seeding processes, crystals of pure gibbsite (aluminium hydroxide) precipitate out. The final stage of the Bayer Process involves another wash before the aluminium hydroxide is roasted. This drives off the watery component of the 'hydroxide' to leave behind a fine white powder of aluminium oxide or *alumina*.

This leads on to the final stage: the Hall–Héroult process. The alumina is dissolved in a bath of molten cryolite or its synthetic equivalent (sodium aluminium fluoride) at a temperature of about 960°C. An electric current is then passed through the electrolyte solution of alumina and cryolite between carbon anodes. The base of the bath acts as a cathode. The current breaks the chemical bond between the aluminium and oxygen in the alumina. This releases the pure metal, which sinks to the bottom of the cell as a layer of molten aluminium to be drawn off. The oxygen released during the electrolysis reacts with the carbon of the anodes to produce carbon dioxide, which bubbles away.

The process takes place on a continuous basis. As the molten aluminium is drawn off, so more electrolyte is poured in. Most of the cryolite or its synthetic equivalent can be reused and recycled. Although this process uses huge amounts of electrical power, historically it gradually brought down the price of aluminium so that by the twentieth century it was no longer regarded as a precious metal. It is still the case, though, that many of the world's aluminium-refining centres are located close to sources of cheap electrical energy. Oil-rich countries and those that can produce hydroelectric power are often favoured as places where aluminium might be produced more cheaply.

Around 60 million metric tonnes of aluminium are produced annually worldwide. Once again, China leads the pack. In 2019, it manufactured a whopping 36 million metric tonnes. Then a long way behind come India, Russia, Canada and the United Arab Emirates, each producing between 3 and 4 million metric tonnes a year. So what is all this aluminium being used for?

* * *

Unlike copper, iron, tin and lead, which have been known about and used for at least a couple of thousand years, aluminium's recent discovery means that it doesn't feature in the archaeological record, at least until Victorian times. Nevertheless, its relative lightness, strength and ductility make it a very attractive and versatile metal for many engineering needs and manufacturing purposes.

Aluminium is reasonably ductile. This means it can be stretched and drawn into thin wires without breaking. It is also malleable and can be hammered, pressed and rolled into thin sheets. Although it is more expensive than iron and steel, its physical and chemical characteristics make it the metal of choice for many products and purposes. Let's consider just a few.

Because aluminium has a density of around one-third that of steel or copper, its lightness gives it a high strength-to-weight ratio. This means that where weight is an important criterion in matters of efficiency and fuel economy, aluminium is often preferred to steel. This has been most obvious in the aircraft construction industry, where less weight means easier lift-off and reduced fuel consumption. In engineering, there is a general push towards finding and using materials that are lighter and stronger, and easily machined, assembled and repaired. Aluminium fits the bill nicely.

In attempts to keep their weight down, some of the very first aeroplane frames were made out of wood covered in a skin of stretched fabric. The Wright brothers, Orville and Wilbur, are generally credited with building the very first successful self-propelled, heavier-than-air flying machine. It took to the skies, just, of North Carolina in December 1903. To keep the weight low, it was made out of spruce, boxwood, waxed twine, steel rods, cotton muslin fabric and an engine specially cast from an alloy of 92 per cent aluminium and 8 per cent copper. Each of the first few test flights lasted less than a minute. The brothers reached heights of little more than 10 feet. But they had done it – the first ever self-powered flight.

From then on, aircraft design developed at a rapid pace. Wood-based aircraft got bigger, went faster and flew higher. In 1941, the British de Havilland aircraft company introduced the 'Mosquito' into the Second World War. It was a twin-engine multi-role combat aircraft whose frame was built almost entirely of wood. It was nicknamed the 'Wooden Wonder' and could reach speeds of over 300 mph.

One of the largest planes ever built was made of birch wood. The Hughes H-4 Hercules, mockingly described as the 'Spruce Goose', was designed as a trans-Atlantic transport aircraft to be used in the Second World War. It was an eight-engined flying boat with a massive wingspan of 321 feet. But it arrived too late to be put into service. It made its one and only flight in 1947. With thirty-six people on board, including its owner Howard Hughes in the cockpit, it flew at a height of 70 feet for a mile, reaching a speed of 135 mph over the waters of Cabrillo Bay, Los Angeles. It is now a museum piece.

The trouble with wood, though, is that it rots. And flying in a rotten aircraft, never mind with a rotten airline, is not to be recommended. Aluminium doesn't rot. In fact, it is very good at resisting corrosion. When exposed to air, a layer of aluminium oxide forms almost instantaneously on the surface. It is this thin layer that prevents any further chemical change. Aluminium oxide is also extremely hard, and although the layer of aluminium oxide that protects the aluminium surface is made up of extremely fine crystals, it is the same stuff of which sapphires are made.

From the early years of aviation, aluminium or one of its alloys became the metal of choice for most aircraft manufacturers. These alloys allowed the development of jet planes and led to cheaper flights. Aluminium and its many alloys are strong, lightweight, workable and relatively inexpensive.

One of the first commercial aircraft to be made out of aluminium was the Ford-Tri-Motor, a three-engined, passenger-carrying plane first introduced in 1926. An earlier bird than Howard Hughes' almost flightless monster, Ford's plane was affectionately known as the 'Tin Goose'. As planes got bigger and faster, constructors found themselves using more and more aluminium to the point where up to 80 per cent of an aircraft's unladen weight was aluminium.

However, today changes are afoot. Many modern aircraft are replacing much of their aluminium content with carbon-fibre composites in which woven mats of carbon are embedded in a plastic polymer resin. Although only half the weight of aluminium, these carbon-fibre composites actually possess greater strength and rigidity than the metal, all of which makes planes even more fuel-efficient and engineers even happier. Their downside is that they are more expensive than aluminium and difficult to recycle. But scientists are up for the challenge. They are working to overcome both of these disadvantages and continue to express optimism that it is only a matter of time before they bring the price down further and work out ways to recycle the plastic polymers.

* * *

In spite of these new developments, aluminium remains an incredibly useful, widely used metal. Its strength is increased when it is alloyed with copper, tin, manganese, zinc or magnesium. In fact, there are over three hundred wrought alloys with at least fifty in common everyday use. As with aircraft manufacturing, the metal's lightness, strength and ability to resist corrosion see it being used in the construction of trains, cars, ships and many other forms of transport.

Aluminium is also a good conductor of energy. From the outset, its high heat conductivity lent itself to making pots and pans. Although not as high as copper, its electrical conductivity, coupled with its much lighter weight, makes it a good choice for carrying electricity along heavy-duty power lines as they string their way from generating stations and wind turbines across the countryside to our towns and cities.

Modern architects not only like the look of aluminium on their buildings, they value its ability to resist corrosion. This makes it virtually maintenance-free. From the builder's point of view, not only does bright, shiny aluminium please the eye, it is easy to fold and bend, cut and weld, so that any desired shape can be achieved.

Two of the first buildings in which aluminium was extensively used are both in New York. Built between 1928 and 1930, the wonderful art deco-styled Chrysler Building, rising over 1,000 feet, has aluminium trim throughout, albeit mainly for decorative purposes.

The Empire State Building, completed in 1931, uses even more aluminium, not just for decoration but also in its fabric. The skyscraper's splendid pointy spire was originally put there as a mooring mast for hydrogen-filled airships, also known as zeppelins, dirigibles or, even more fondly, 'blimps'. Only a couple of hazardous attempts were made to hover above the mast, however, and no successful moorings were ever made. Experts said that the whole idea of docking these behemoths of the sky to the top of a New York skyscraper was never going to work and so the idea was quietly dropped.

The sinuous, wavy roof of Zaha Hadid's London Aquatics Centre, originally built for the London 2012 Olympic Games, is tiled with flowing aluminium panels. And if it's mad, fun and funky you're after, look no further than Frank Gehry's shiny aluminium-clad Luma Arles tower as it twists and twirls skywards. The irregular look of the tower is supposed to reflect the craggy rock formations that rise nearby the southern French town of Arles, where Vincent Van Gogh painted his equally vibrant and colourful *Café Terrace by Night* in 1888.

* * *

More immediate uses of aluminium can be seen in the make-up of smartphones, tablets, laptops and flat-screen televisions. Whenever you open a shiny, silver laptop or hold a smooth, sleek smartphone, it could well be aluminium you're fondling. Aluminium makes our high-tech gadgets look smart and sophisticated while at the same time making them feel light and durable. By weight, some smartphones are 24 per cent aluminium.

From chairs to tables, picture frames to window frames, curtain rails to lamps, our houses are full of things made out of aluminium. But perhaps the most taken-for-granted use of aluminium is when it's rolled out into a thin foil. In particular, the food industry has taken to aluminium foil in a big way.

Foil is made by passing sheets of aluminium between two rotating rollers under pressure. Depending on its ultimate use, the foil can be made of pure aluminium or one of its many alloys. It can also be rolled to a variety of thicknesses, from heavy-duty foils at around 0.24 mm thick to incredibly thin foils a mere 0.004 mm thick. We're all familiar with kitchen foil and its amazing usefulness. You can pack sandwiches in it, wrap fish in it before baking it in the oven, cover leftover food with it before placing the container in the fridge, and grill things on it, preventing mess and drips falling onto the tray below.

However, where you most often meet foil in both variety and quantity is in shops and supermarkets. Anything hermetically sealed with an aluminium lid or packaged in an aluminium sleeve will stay fresh and have an extended shelf life well beyond anything that was possible before the foil's widespread use. Think lids on milk bottles, yoghurt pots, tubs of margarine and even plastic-lined aluminium screw caps on bottles of wine, whisky, olive oil, vinegar and fruit juices.

Medicines, too, are often packaged in aluminium foil. Pills and tablets come in aluminium foil blister packs. Capsules, potions and lotions can be sealed beneath aluminium caps. Aluminium foil offers complete protection against light, moisture, oxygen and other gases. It stops microorganisms and bacteria sneaking in. That's very useful and good to know.

And when you next finish running a marathon, you may well be given a shiny, highly reflective blanket of very thin foil to wrap around your tired body. These blankets help regulate your body temperature, which tends to drop drastically once you stop running. They are known as 'space blankets'. NASA first developed the technology when it was figuring out ways to keep heat out of sun-drenched satellites. But turn the blanket around and it will be just as effective at keeping the heat in. The blankets are made by vacuum-depositing a very thin layer of pure aluminium vapour onto a very thin, durable plastic film substrate. Their use is now widespread, not just for keeping athletes warm but also for mountaineers, polar explorers, patients in danger of suffering hypothermia, and in houses.

One of the biggest uses of aluminium is in the manufacture of cans, especially for drinks, both alcoholic and non-alcoholic. The idea of preserving food in a sealed can was first developed early in the nineteenth century. Steel and tin-plated steel were used to fabricate the cans and, even to this day, many people still refer to cans, however made, as tin cans. Although steel cans are still manufactured in huge quantities for the food industry, preserving anything from baked beans to rice pudding, aluminium cans dominate the drinks market. Around 180 billion aluminium drink cans are made each year. That's roughly twenty-three cans for every man, woman and child on the planet, which of course means that given that many of us never even open a drinks can, there must be others snapping their ring pulls hundreds of times a year.

It wasn't until the late 1950s that cans made of aluminium first made their appearance. The virtues of aluminium cans as non-corrosive, lightweight, recyclable containers were quickly appreciated. In no time at all, the aluminium drinks can conquered the world – especially after the ring-pull was invented.

Inevitably, manufacturing companies got smarter and smarter in their technological know-how. Cans are pressed from an aluminium alloy that is 97.25 per cent aluminium, with the remainder being made up of magnesium (1 per cent), manganese (1 per cent), iron (0.4 per cent), silicon (0.2 per cent) and copper (0.15 per cent). A thin, safe plastic lacquer veneer coats the inside. The can is then filled with the drink before the tops, having been made separately, are finally welded to the neck.

Incredibly, modern 0.33 litre cans weigh less than 20 grams when empty. Their walls are no more than 0.08 mm thick, according to the International Aluminium Institute. Although the costs of manufacturing are still marginally dearer than those of glass, the lighter aluminium cans are much cheaper to transport. And they don't break.

The other plus, of course, is that aluminium cans, in theory, can be endlessly recycled. Making cans from recycled aluminium requires 95 per cent less energy than producing the metal from its original ore. The rate of recycling does, of course, depend on users being tidy-minded and environmentally conscious. However, counting the number of discarded empty cans littering city streets, motorway verges and country hedgerows can be a dispiriting experience. There appears to be a population of bevvy-swilling bozos who, wishing to show their strength by crushing flat their drained cans and then tossing them out of their car windows, lend support to the Yorkshire proverb, 'Strong int'arm and thick in't head'.

* * *

Even though it took a long time, aluminium, from being the most abundant metal in the Earth's crust, has finally hauled its way up the rankings to be the second-most produced metal in the world after iron and steel. Although iron and steel dwarf all other metals, with more than 1,800 million metric tonnes produced annually worldwide, more than 60 million metric tonnes of aluminium are poured, cast and rolled out of refineries every year. Considering copper, tin, iron and lead have been around for a couple of millennia or more, that's some achievement in less than two hundred years.

Unlike iron and copper, however, there is little metaphorical or idiomatic use for aluminium, unless, of course, we count cans as aluminium proxies. Decisions can be delayed when we kick a can down the road. We can be held responsible when we're asked to carry the can. And when we've successfully completed whatever we've been doing, all we have to do is save it and keep it safe. This is what movie-makers did when their filming and editing was done and all they had to do was pack their light-sensitive reels into airtight aluminium cans to be shown in cinemas up and down the land. The movie was in the can.

Chapter 9

Plastics

As with concrete, plastic doesn't always enjoy the best of reputations. 'It's only plastic' is the disappointed response when someone discovers that a flower is artificial, a toy has broken, or what they thought was a smart leather bag isn't leather at all. The enormous amount of plastic that is polluting the oceans and getting into the food chain is of huge concern. It is plastic's ubiquity, dissembling qualities and cheapness that have seen it enter our everyday language. 'It's not real; it's only plastic.' Or 'Go on, flash the plastic', when you're being encouraged to spend wildly using your credit card, made of plastic.

But before we take a longer look at the problems with plastic, we really do need to recognise what an extraordinary, versatile, innovative, durable material it is. The clue to its versatility comes from its name, 'plastic', meaning pliable, easily moulded and readily shaped.

The French literary critic Roland Barthes saw plastic as a substance capable of 'infinite transformation … it is less a thing than the trace of a movement'. It is, he writes, a quick-change artist but also 'a disgraced material' lacking the integrity of minerals and metals as it shape-shifts into anything and everything. 'Despite having names of Greek shepherds (Polystyrene, Polyvinyl, Polyethylene) … plastic is in essence the stuff of alchemy.'

Compared to most of the other stuff we've looked at, plastic is the new kid on the block. It didn't seriously arrive until the middle of the twentieth century, since when its manufacture and use have been unstoppable. It's hard to imagine how we could have lived without it. In the material world, this is the Age of Plastic.

Plastics are everywhere, in everything, on everything. Pots for yoghurt, humus, cream, soups. Flimsy single-use shopping bags made of polyethylene. Bottles for milk, fruit juices, soft drinks, water, tomato ketchup, olive oil, shampoos, medicines. Plastic trays for fish, meat and vegetables all wrapped in plastic clingfilm. Toothbrushes, toilet seats, car fabrics, door handles and light switches. Clothes made from polyesters and nylon. Vinyl records. Plastic casings for computers, phones, televisions. Plastics for hip joints and knee replacements, surgical gloves and bedpans. Plastics in synthetic car tyres and paints. If you're sitting indoors while reading this, look around you and you'll be hard pressed not to see at least a dozen things made out of some kind of plastic. The list just goes on and on.

Millions and millions of tons of plastic are made every year. Most of it is made from petroleum and the by-products of the oil industry. To begin the story of plastics, we therefore have to start with how oil is formed, where it accumulates and how it is extracted. Oil is yet another example of how biology and geology interact.

* * *

Surprisingly, perhaps, carbon is slightly more abundant in the universe (0.5 per cent) than in the Earth's crust, where it makes up about 0.18 per cent. Carbon is the element of life. Because of carbon's fascinating chemistry, it can combine with itself, as well as other elements, especially hydrogen, to form thousands and thousands of different hydrocarbon molecules. It is when these hydrocarbons configure themselves into self-replicating, highly complex molecular entities that we have life, as Darwin said, in 'endless forms most beautiful'.

Oil is produced when organic biological matter dies, sinks, gets buried and undergoes chemical change. The most common sources of biological matter that ends up as oil include microscopic single-celled plants (known as diatoms), protozoans and larvae.

If the waters in which these plants and animals thrived were rich in such life, the second condition for oil to form is that when they die and rain down on swamp bottoms, lake beds or sea floors, oxygen levels are low or absent. This means that the hydrocarbons can't oxidise and bubble away as carbon dioxide. It also means than other life forms can't live there either. There can be no other creatures to scavenge the dead remains. The organic matter thus settles and slowly accumulates along with the muds, silts and sands that gently drizzle down, layer upon layer, over thousands and thousands of years.

As the organically saturated rocks sink and become buried deeper and deeper, they are heated and compressed. This is the start of the chemical changes that convert dead organic matter into crude oil. The conversion takes place at between 50 and 200°C, though 100 to 120°C provides the ideal condition, a Goldilocks zone known as 'the oil window'. Too cool and oil doesn't form. Too hot and it boils away to form methane and other natural gases.

The organically rich shales, sandstones and limestones in which the oil first forms are known as *source rocks*. The oil and organic matter colour these rocks a dark, dirty brown or black. The exact nature of the particular crude oil formed depends on the composition of the original organic matter and the physical conditions under which it was converted. Those that contain detectable amounts of hydrogen sulphide (the smell of rotten eggs) are known as 'sour crudes'. Oil refineries prefer the less smelly, low-sulphur oils, which they lovingly call 'sweet crudes'.

However, whatever the oil's eventual make-up, it's wonderful to remember that the energy it contains is sunlight, sunlight that was once captured and converted into life through photosynthesis, then trapped and transformed, before finally being freed and released once more by humankind's ingenuity.

Oil, or petroleum, can therefore be thought of as rock (Latin, *petra*) oil (Latin, *oleum*). It is typically a mixture of various gas, liquid and solid hydrocarbons. Most crude oils contain a number of the following: methane, ethane, propane, butane, pentane, benzene and asphalt. Occasionally, the oils seep to the surface. This is the origin of bitumen, tar pits and tar sands.

Being predominantly liquid and gas, once formed underground and under pressure, the oil has a tendency to seep upwards until it reaches a rock from which it cannot escape. This is known as a *reservoir rock*. The reason the oil becomes trapped is that the reservoir rock, which *is* porous, is capped or surrounded by rocks that are not porous. These are *caprocks*. They typically include lithified clays and other very fine, compacted rocks. In practice, there are a number of geological structures that lend themselves particularly well to oil entrapment and accumulation. Rocks that are bent and buckled upwards into the shape of arched anticlines, and salt domes that push their way up into non-porous caprocks, are particularly effective structures for trapping oil.

In economically worthwhile deposits, anything over a gallon of oil can be found for each cubic foot of rock. The unit of oil volume is known as a 'barrel'. It is equal to 42 gallons.

Although oils that had leaked to the surface had been known about since ancient times, it wasn't until the nineteenth century and the burgeoning science of geology that it was suspected that oil deposits, like coal, might well be found deep under the ground. The first successful oil well was drilled in Titusville, Pennsylvania, in 1859. Even today, with vastly improved techniques and fancy equipment, geologists can never be absolutely sure that when they mark a spot for drilling, oil will actually be found. However, when the drill is successful and oil is found for the first time in a new field, the well is described as a 'wildcat well'.

Natural conditions sometimes mean that when the drill pierces a reservoir rock in which the oil is under tremendous pressure, the oil and gas will erupt like champagne out of a bottle and spurt high into the air. Until controlled, this is known as a 'gusher'. In most cases, however, the oil has to be pumped out.

The world's leading oil producers in 2020 include the USA, Saudi Arabia, Russia, Canada, China, Iraq, the United Arab Emirates, Brazil, Iran and Kuwait, although market prices and politics affect who is pumping the most out in any one year. This picture changes when oil reserves are considered. Topping the league of who has the most petroleum still left in the ground are Venezuela, Saudi Arabia, Canada, Iran, Iraq and Kuwait.

In Britain, small amounts of on-shore oil deposits have been found. Spurred on by the need to find fuel to support the military forces in both the First and Second World Wars, oil was found at Hardstoft, Derbyshire, and a few years later, in 1939, at Eakring in Nottinghamshire.

The first wells at Hardstoft drilled through the Carboniferous Coal Measures, and on through the underlying Millstone Grit and Limestone shales. On the night of 27 May 1919, oil was struck at a depth of 3,070 feet in the top layers of the Carboniferous Limestone. The first oil to flow from the well burbled out on 7 June 1919 at the rate of eleven barrels a day! Over the next eight years it produced about 20,000 barrels before production finally ceased in 1927. Nevertheless, its modest success means that it can claim to have been Britain's first oil well.

Since Hardstoft, a number of small oil fields have been found on Britain's mainland, including parts of the East Midlands, first drilled at the beginning of the Second World War. Wytch Farm in Dorset still operates today, though rates of production are well down on the peak levels reached back in the late 1990s when over 100,000 barrels a day were pumped to the surface. It still remains Europe's largest on-shore oil field, although it is dwarfed by the oil and gas found and extracted from the North Sea.

Although it had been suspected for some years that the rocks beneath the North Sea might contain oil and gas, it was not until the late 1960s and early 1970s that significant amounts were discovered. In 1999, peak oil production was reached at around six million barrels a day. Now in slow decline, the combined reserves of the Norwegian and British sectors still place the region (in 2020) around seventeenth in the league table of world producers.

★ ★ ★

Most of the world's oil and gas ends up as fuel of one kind or another to propel cars, trundle lorries, fly aeroplanes, drive ships and fire power stations.

Crude oil itself has to be processed before it can be sorted into its various components. The oil extracted is a mixture of hydrocarbons. Each hydrocarbon has its own molecular weight, chemical properties and physical characteristics. These can be sorted and separated from each other by a process of 'fractional distillation', otherwise known as oil refining.

The oil is first heated to a vapour. It is then passed upwards through a distillation tower. At different heights up the tower there are 'trays'. The vapours are very hot at the bottom of the column but become cooler as they rise so that different fractions condense in the trays at different heights. The lighter the fraction, the higher up the tower it condenses. In an average crude oil, the fractions that can be separated off into the trays, beginning with the lightest, include dissolved gases, petroleum ether, petrol and gasoline, naphtha, kerosene (paraffin), diesel and gas oils, lubricating oils, fuel oils and, heaviest of all, asphalt (bitumen).

Only around 4 per cent of all the oil produced each year is turned into plastic. Further refining is necessary before the raw ingredients to make plastics can be obtained. Refineries employ the technique of thermal 'cracking'. This process uses heat to break down some of the more complex hydro-carbons into simpler chemical molecules. Many of these are of particular interest to the plastics industry. They include ethane (C_2H_6), propane (C_3H_8), butane (C_4H_{10}) and benzene (C_6H_6).

Then, more heat and chemical wizardry conducted on these raw materials eventually leads to the production of long chemical chains in which carbon and hydrogen mol-ecules become linked to form polymers. Ethylene and propylene are two examples. When melted, cooled and cut into pellets, the polymers are ready for use by the plastics industry.

★ ★ ★

The history of plastics goes back to the mid-nineteenth century. In 1869, the American inventor John Wesley Hyatt produced a plastic by chemically treating a plant cellulose derived from tree pulp with camphor, an organic oil. He called it celluloid. Its initial success saw it being used to make spectacle frames, combs, dentures, buttons and bil-liard balls. Up until then, billiard balls had been made of ivory. Manufacturers of the new 'plastic' made a virtue of celluloid's ability to replace ivory and so helped save the elephant from the greed of hunters and poachers. Celluloid was also used to make photographic film. This ushered in the beginning of a cultural revolution. People began to go to the 'cinema' to see moving pictures, where strips of translucent celluloid whirled around reels and zipped past a bright light that projected the pictures of each whirring frame onto a large silver screen.

Plant cellulose is a natural polymer. The term 'polymer' – meaning 'having many parts' – started to be used to describe synthetic materials that were also built up of many parts, usually in the form of very long molecular chains arranged in repeating units. Carbon atoms, typically bonded with atoms of oxygen, hydrogen and nitrogen, form the backbone of most of these natural polymers. The great length of the carbon chains makes them strong, lightweight and flexible. It is these physical properties that allow them to be shaped and moulded, to be plastic.

Hyatt's achievement was revolutionary. Most manufactured products had to work with and around the constraints of natural materials such as wood, metals and stone. Not only could plastic mimic the characteristics of natural materials, it opened up the prospect of creating materials with new properties and unimagined uses. And it was cheap! A plastics revolution was on the way.

Bakelite was the first fully synthetic plastic. No plants were harmed in the making of this product. Leo Baekeland, a Belgian chemist who had moved to America, had been looking for a synthetic substitute for shellac, a natural resin used to insulate electrical wires and circuits. Working with phenols and formaldehyde, in 1907 he eventually came up with – deep breath – polyoxybenzylmethylenglycolanhydride. Clearly with a chemical moniker like that, the plastic needed an altogether friendlier name, and so in honour of its inventor it became known as Bakelite. It was certainly a good insulator. It was also durable and heat-resistant, and could be moulded into any shape.

Bakelite was marketed as the 'material of a thousand uses'. In no time at all it was being fashioned into radio casings, cigarette boxes and, most iconic of all, the Bakelite telephone.

The early successes of celluloid and Bakelite encouraged the chemical industry to develop new polymers. More and more plastics began to be invented. What to do with them wasn't always clear, but time and time again, a new plastic was found to have surprising and unexpected properties. It was then left to the inventiveness of scientists, engineers and entrepreneurs to come up with possible uses and applications that could not even have been imagined until the new plastic had been created.

By the early 1930s, acrylic sheets or Plexiglasses were being used as an alternative to glass, especially in the manufacture of aircraft windows. In 1935, Wallace Carothers invented a synthetic fibre. He called it 'Nylon' – a thermoplastic silky material. In no time at all, it was appearing in toothbrush bristles, synthetic silks for parachutes, rope fibres, carpets and, of course, nylon stockings. 'In product after product, market after market,' says the science writer Susan Freinkel, 'plastics challenged traditional materials and won, taking the place of steel in cars, paper and glass in packaging, and wood in furniture.' The possibilities of plastics seemed endless. But then another breakthrough occurred.

As we've seen, a number of liquids and gases appear as by-products of refining oil. In the 1930s, using high pressures and catalysts, British scientists found ways of turning some of these by-products, including ethylene, benzaldehyde and a few other compounds, into long-chain polymers. The result became known as *polyethylene*, meaning lots of ethylene molecules linked together. These days, after the polyethylene liquid resin is cooled, it is first stretched and then cut into pellets. It is then ready for use in one or other of the vast number of injection moulds used by the plastics industry to shape everything from plastic bottles to food packaging, toys to toothbrushes, buckets to bowls.

Chemists then turned their attentions to another crude oil by-product. Propene, also known as propylene, is a colourless gas (C_3H_6). Again, under heat and pressure and in the presence of a catalyst, in 1954 scientists learned to transform the gas into *polypropylene*. It can be made into yoghurt pots, carpets, disposable nappies, car interiors and medical equipment.

And still the alchemy continued. Other elements began to be introduced into the make-up of the long-chain polymers. The addition of chlorine produced *polyvinyl chloride* or PVC to make shower curtains, water pipes and building panels. Introduce fluorine and you get *polytetrafluoroethylene*, or Teflon, with its wonderful non-stick qualities so good for pans, low-friction bearings and slippery politicians.

Polystyrene, and expanded polystyrene (puffed up with air into a 'synthetic meringue'), is used for insulation, packaging around fragile goods, and to make coffee cups. *Polyurethane* produces soft, squishy cushions, comfy car seats, running shoes and Lycra. It can also be used to fabricate the plastic interiors of fridges and chiller cabinets, as well as being added to varnish. Unsaturated polyesters are shaped to form boat hulls, bathtubs, shower stalls and garden chairs. The chemical conniving of these industrial scientists seemed to know no limits.

* * *

In 2019, the world produced 368 million metric tonnes of plastic. Or a billion kilograms each day. Four times as much as was made in 1989. We can't seem to live without it, even as it threatens to choke our rivers and pollute our seas. The very properties that make plastic so attractive to producers and consumers are also the reasons why it has become a global problem. Plastics are durable. Most resist chemical breakdown. Even if they physically disintegrate into smaller pieces, the bits hang around for decades, for centuries, for millennia.

Take single-use plastic bags. In technical terms they're a marvel. They're very thin, yet strong, weighing next to nothing but capable of carrying up to a thousand times their own weight. They're cheap, waterproof and convenient. But although they might tear and rip, still they persist as their tattered remains festoon the highways and byways, 'witches' knickers' clinging to bushes and fences.

Similar engineering miracles are witnessed in the creation of the humble plastic bottle. Firm, strong, lightweight, shatterproof and easily carried. Being so much lighter than glass bottles, they are much cheaper to transport from factory to shop. But once their contents have been drunk, the temptation is to throw them in the bin or, if you're a litter lout, over your shoulder or out of your car window, where the bottles lie beneath the witches' knickers, creating a distressed world of battered bottles beneath bedraggled blossoms of tattered plastic. They do have one saving grace, though. The *polyethylene terephthalate* (PET) from which the bottles are made can be recycled to make a new generation of plastics, from polyester fibres to carpets.

However, cycles have a habit of coming round. Back in the nineteenth century, the first plastics were made from plant cellulose, which does degrade. There are now attempts to rediscover the virtues of plant-based plastics. For example, the Danish company Lego has for years made its billions of toy bricks out of crude oil-based plastic known as acrylonitrile butadiene styrene. But it is a company with the beginnings of an environmental conscience. In 2018, the Lego Group announced that it would be making a small number of its pieces from sustainable, plant-based plastics, sourced from sugar cane. Fittingly, the first things to be made out of this botanically derived material are plastic trees, bushes and leaves. The new-style Lego products are still being made from polyethylene, but one derived from ethanol extracted from the sugar cane. The company claims that the bioplastic can be recycled many times, although it is still not 100 per cent biodegradable.

We'll revisit plastics in the concluding chapters, where its polluting vices are beginning to outweigh its versatile virtues.

Chapter 10

Lithium, Rare Earths and the Information Age

Every time you touch or swipe the screen of your smartphone or tablet, it is a thin, ultra-sensitive, electrically conductive layer of indium tin oxide coating the glass that is responding to the movement of your finger. Turn on your high-definition TV and be dazzled by the intensity of the colours. Yttrium, terbium and europium are the elements that are used to make the materials that produce this visual vibrancy. Indium, yttrium, terbium and europium are just four of the fifteen rare earth elements that play a vital role in our modern high-tech, digital world.

The names of most of the elements and minerals associated with the rare earths remain relatively unfamiliar. However, over the past few decades, these elements have played a subtle but critical role in the revolution that is the Age of Information and Technology. They crop up in all kinds of devices. They are found in batteries and magnets, computer hard drives and phone screens, cameras and medical scanners. The modern world has become highly dependent on these elements and their minerals, so much so that their geological origins, restricted geographical distribution and the challenges associated with their economic extraction have become matters of increasing political sensitivity.

Let's begin with lithium (Li). It wasn't fully isolated and identified as an element until 1817. The Swedish chemist Johan August Arfvedson recognised it as a new metal, similar to sodium but lighter. He realised this was a new alkali metal and a lighter version of sodium. However, unlike sodium, he was not able to separate it by electrolysis. It was not until 1855 that the German chemist Robert Bunsen (he of the burner fame) and the British chemist Augustus Matthiessen obtained it in bulk through the electrolysis of molten lithium chloride.

With only three protons in its nucleus, its atomic number is three, the third element in the periodic table after hydrogen (atomic number 1) and helium (atomic number 2).

Silvery-grey in colour, it is the lightest of all the solid elements. It will float on water but easily reacts to form lithium hydroxide. It is soft. You can cut it with a knife. Like its alkali cousins, sodium and potassium, it is chemically very reactive and never occurs freely in nature. In oxygen, it burns with a beautiful carmine red flame. Crucially it has very good electrical conductivity, too, which makes it ideal for use in batteries.

* * *

Lithium is sparsely present in many rock types. Its abundance in the Earth's crust barely reaches 20 parts per million. That's hardly enough to make its extraction economically feasible. However, as with all of the other elements that we have learned to use and exploit, nature gives us a helping hand. Several geological processes have led to its relative concentration in various parts of the world.

Granites and similar igneous rocks generally have higher concentrations of lithium minerals in their make-up than other rocks. We have already seen that molten granitic magmas form beneath newly rising mountains where one tectonic plate meets another. Lying deep below the surface, granites slowly cool. This allows the crystals to grow large. Depending on their physical properties, some minerals will begin to crystallise before others. Heavier minerals will sink while lighter ones will rise, and so further concentrations of particular minerals can occur at different levels and depths within the molten granite.

There is a particular rock known as a *pegmatite* that often forms around and above large magmatic intrusions, including those of granite. Pegmatites occur as the more fluid and volatile minerals of the magma rise, pierce and inject themselves into the rocks around and above the intrusion. Here, they form veins, large lens-like sheets, and swarms of igneous dykes. The chemical characteristics coupled with the slow cooling of these pegmatitic intrusions allow very large crystals to grow. These can measure several inches in length. Large, coarse pink and grey crystals of feldspar, white quartz and scintillating micas dominate the rock's mineralogy. But other minerals also cool and grow, although which particular examples depends on the chemical character and cooling history of the parent granite.

Although pegmatites are fairly common, those bearing minerals of lithium are relatively rare. Until fairly recently, it was these lithium-bearing pegmatites, and the clays that form when the rocks weather, that supplied most of the world's lithium ore. The most common lithium-rich minerals in these lithium aluminium silicate ores are spodumene, lepidolite and petalite.

The three largest commercial lithium-bearing pegmatite deposits are found in North Carolina, Manono in the Democratic Republic of the Congo, and Greenbushes in Western Australia.

It is the huge open-pit and underground mines of the Greenbushes area in south-western Australia that currently supply around 40 per cent of the world's annual lithium. This helps make Australia the leading lithium producer, for the moment. The ore occurs in ancient Pre-Cambrian rocks, 2,525 million years old, half the age of the Earth itself. After mining, the ores are subjected to a number of physical and chemical processes at the end of which industrially useful lithium oxide concentrates are achieved ready for export and use.

Mining for lithium in hard rocks, such a granitic pegmatites, is expensive. From digging up the ore to achieving industrially useful concentrations of lithium oxides and carbonates takes a lot of crushing, grinding, dissolving, separating, filtering, roasting and chemically treating. Moreover, the whole operation uses a huge amount of energy and involves a major assault on the landscape.

Although there are a few sedimentary deposits of lithium ores, the up-and-coming alternative to mining obdurate igneous rocks is to find brines relatively rich in lithium salts and evaporate them. Brines describe any fluid, especially water, in which there are dissolved solids. The most common dissolved solid in seawater is sodium chloride (common salt).

Together, sodium and chlorine make up 90 per cent of all of the dissolved ions in the oceans, the briny deeps. On average there are 35 grams of salt per litre of sea water. If you evaporated the waters of all of the world's seas, you'd be left with a salty layer about 130 metres thick. In that layer, as well as sodium chloride, potassium chloride and calcium carbonate, very, very small amounts of the salts of lithium will occur.

Rivers and streams pour these dissolved salts into the oceans. They are the end result of all the weathering that takes place on the land. As rain falls through the atmosphere, it dissolves tiny amounts of carbon dioxide. In effect, it becomes a very weak solution of carbonic acid. The ever-so-slightly acidic rain has the power to dissolve, over time, many of the minerals that make up the rocks of the land. Rivers then take these dissolved salts out to sea, the salty sea.

However, it is also possible for rainwaters and volcanic geothermal waters to dissolve rock minerals that don't get washed out to sea but accumulate on land, underground or as salty inland lakes. If these waters happen to percolate through and around lithium-rich rocks, such as the igneous pegmatites we mentioned above, then lithium salts, along with others, will be leached and scavenged. The resulting brines might have enough lithium in them to make their commercial exploitation worthwhile, especially if they end up in salt flats, the Spanish for which is *salar*. Lithium concentrations in some salt flats can reach as high as 1,000 to 4,000 parts per million. And this is what we find in a number of the world's salt flats, including Salar de Atacama in Chile, Salar de Hombre Muerto in Argentina, and Zabuye Salt Lake, high in the Tibetan Plateau.

The largest known reserves of lithium are found high up in the Bolivian Andes, a fairly young mountain chain that was pushed up around 50 million years ago as the Pacific plate subducted beneath the South American plate. Like all plate-edge mountain chains, they have hearts of granite and intrusions of pegmatite.

Salar de Uyuni is the world's largest salt flat, deep beneath which lie magmatic rocks. Hot, geothermal waters boil their way through these igneous intrusions, leaching ions of sodium, potassium, boron and lithium as they do so. The salar was formed tens of thousands of years ago when several prehistoric lakes merged. The salty lake waters began to evaporate in the dry air and, as they did so, the various alkaline salts began to crystallise and precipitate. Today, the flats are covered by a few metres of salt crust, which can itself be mined for commercial use. But below the crust is a brine rich in lithium.

In order to achieve sufficiently high and economically profitable levels of lithium concentration, the brines have to undergo further stages of evaporation. First, the brines are passed through a series of ponds, where, in these high, dry altitudes, the sun evaporates the water. Lithium salts are more soluble than most other salts. Those of sodium chloride, potassium chloride and calcium sulphate are the first to precipitate and so they are harvested first. This leaves the remaining brine increasingly rich in lithium compounds. There are other salts, too, including those of magnesium and boron. These also have to be removed. Eventually, lithium chloride is left, either in a concentrated brine or as a precipitate. The lithium-rich brine is then pumped to a recovery plant, where sodium carbonate is added. This converts the lithium chloride into a lithium carbonate slurry, which is filtered and dried.

Each briny deposit has its own unique characteristics, necessitating slightly different physical and chemical processes, but in all cases the end result is an almost pure lithium compound, usually lithium carbonate, although some lithium hydroxide and lithium chloride can also remain. The compounds are then electrolysed to produce the pure lithium metal.

Cornwall is famous for its tin mines. Over two thousand years ago, the Romans were already importing tin from the county. In later centuries, copper was also found and mined. Sadly, the last tin mine in Europe, South Crofty, finally shut in 1998, although there is talk of reopening it in the near future. However, there is the more immediate prospect of another metal being mined in Cornwall. High-grade, lithium-rich brines have been found in the county's hot, geothermal waters. The metal-rich waters originate in the granites that lie beneath the south-west peninsula, from the Isles of Scilly to Cornwall and Devon. And as a bonus, the hot brines can also be used to create electricity, not only to power the lithium extraction and concentration processes, but to provide a source of energy, too. A new metal mining era beckons for Cornwall.

Lithium production worldwide has risen dramatically over the past twenty years. The uses to which lithium and its compounds can be put grows year on year.

* * *

One of the earliest uses of lithium oxide was as a flux added to glass and ceramics. This reduces their melting point and viscosity, saving energy and cutting costs. It can also help increase hardness and shine.

Although lithium salts, including lithium carbonate, have been used in medicine for some time, it was not until 1949 that they were found to be effective in stabilising the moods of people diagnosed with various affective disorders, including bipolar disorder. 'Lithium', in the form of one or other of its salts, is now a widely prescribed drug in psychiatric medicine.

However, the biggest use of lithium today is in the manufacture of batteries, both non-rechargeable and rechargeable, big and small, disposable and reusable. In small, non-rechargeable, disposable batteries, the lithium compound is used for the anode. Lithium batteries have a longer life and so tend to be used in such things as heart pacemakers, watches, cameras, clocks and cordless power tools. In a typical mobile phone, the battery will contain about 2 or 3 grams of 'lithium carbonate equivalent' (LCE), and the smarter the phone, the more lithium the battery will need. A laptop computer will require a lithium-ion battery that contains up to 30 to 40 grams of LCE.

Rechargeable batteries, or 'lithium-ion' batteries, have lithium in the electrolyte. These batteries can be found in electric and hybrid cars, mobile phones, laptop computers, cameras, aeroplanes and satellites. In these batteries, during use, lithium ions move from the negative electrode through the electrolyte to the positive electrode. The ions move in the opposite direction when the battery is being recharged.

There are many upsides to lithium–ion batteries, but they are sensitive to the cold. If you've ever tried to take a photo with your smartphone on a cold day, up a mountain, with the wind blowing, you may well have found that the camera function doesn't respond. Before it works, the camera is likely to need a warm cuddle.

With worries about oil production and carbon dioxide emissions, we are seeing a massive growth and investment in electrically driven cars, vans, lorries, boats and possibly even planes, whether hybrid or all-electric. They all need very large batteries and therefore a huge lithium content. An all-electric vehicle, depending on its size, might need a battery in which there is as much as 8 to 40 kilograms of lithium carbonate equivalent.

As with all types of technological progress, there are environmental swings and roundabouts. Electric cars are good because they don't produce greenhouse gases and so help combat climate change. But their batteries use a huge amount of lithium, whose ores are found in only a few countries, most notably Chile, Argentina, the USA, Australia, China and Bolivia. Extracting the metal has an environmental cost. In the case of lithium-rich brines, local, highly complex hydrological and ecological systems get disturbed. Mining for lithium in hard rocks, often open-cast, destroys landscapes.

Sources vary, but the five biggest importers of lithium carbonates and oxides are currently believed to be Japan, South Korea, China, Europe and the USA. Given the world's increasing dependence on lithium, this makes the metal and the production of its ores a politically sensitive business.

In its anxiety not to be dependent on countries outside of Europe, considerable interest has been taken in Portugal's potentially lucrative lithium deposits. The Portuguese are calling it *petróleo branco*, 'white oil'. The mining companies and their government supporters argue that mining the metal will help play a significant part in reducing carbon emissions. Environmentalists point out that hillsides are being ripped open and local ecologies destroyed. Even mining lithium and manufacturing electric vehicles produces huge amounts of carbon dioxide, a factor in the carbon-reduction equation that often gets forgotten.

* * *

Similar stories can be told for the rare earth elements (often abbreviated to REE).

Most accounts of the rare earth elements begin by telling us that, actually, many of them are not that rare. The general abundance of all rare earths in the crust is around 200 parts per million. Of the individual rare earths, cerium is the most abundant at about 60 parts per million and thulium the least abundant at around 0.4 parts per million. These averages might not seem very high, but remember, much more familiar elements fail to reach these crustal concentrations. Lead, for example, occurs at only 11 parts per million, while silver occurs at a mere 0.1 parts per million. The rarest rare earth, thulium, is actually 125 times more common than gold, while cerium at the other end of the rare earth scale is 15,000 times more abundant than gold.

The description 'rare' is partly based on their late discovery and the difficulties that scientists had in chemically separating and isolating them for investigation. The description 'earth' refers to their diffuse, dispersed presence in rocks and the rarity of the minerals in which they can be found. Geologically speaking, they are seldomly sufficiently concentrated to make their presence obvious or easy to mine. It wasn't even until the twentieth century that all of the REE were identified.

The REE are sometimes referred to as the rare earth *metals* as they are similar in many physical and chemical respects to more familiar metals. Together, the REE form a group of fifteen physically and chemically similar elements. Because of their interesting physical and chemical properties, these elements lie in a separate block within the periodic table. Their atomic numbers range from 57 to 71. Together they are known as the 'lanthanides'. Two extra non-lanthanide elements with similar chemical and physical properties are usually added to this 'rare earth' group: scandium and yttrium.

Although they don't occur in their pure form in nature, when isolated, the REE take on a lustrous, grey to silver appearance. They are soft, malleable and good conductors of heat and electricity. They possess some interesting and unusual magnetic, electrochemical and optical properties. It is these factors that have led to their increasing use in the electronics industry and the key role they are playing in the digital and information age.

Although the REE occur in very small amounts in a variety of rock minerals, their presence occasionally reaches a worthwhile level. The three most common minerals in which there might be a detectable REE presence are bastnäsite, monazite and xenotime.

As we've seen with so many other elements and minerals, before the REE can reach anything like an economically worthwhile presence, geological processes have to occur to increase their concentrations. Sometimes this is achieved deep within igneous magma bodies and through hot, mineral-rich hydrothermal fluids flowing out from the magmas into the surrounding rocks. These are known as primary deposits. The most commercially attractive deposits tend to be associated with these primary, magmatic processes, particularly around types of alkaline and silica-poor igneous rocks known as carbonatites (which, as the name implies, are rich in carbonate minerals).

One place where carbonatites can be found is the site of large-scale faulting and rifts in otherwise stable continental shields. Here, deep plumes of alkaline magma slowly rise from the upper reaches of the mantle, making their way towards the surface along these lines of crustal weakness. Under pressure from the rising plume, the crust begins to swell and stretch, before finally rifting apart. Good examples can be seen in the East African Rift Valley and the ancient shields of eastern Canada. However, two of the world's largest REE mining operations are found in ancient Pre-Cambrian rocks in China. At the Bayan Obo mining district in inner Mongolia, the REE ores appear in rocks over a billion years old. Farther south in Sichuan province are the Maoniuping mines, sited in the rifts of a carbonatite-syenite rock complex.

Concentrations can also be achieved when these igneous rocks and mineral veins are subjected to erosion and weathering. These are known as secondary deposits. They typically occur in river sands and gravels, clays and evaporites, whether geologically ancient or new. Other secondary deposits have been found on beaches or even the sea floor.

* * *

But remember, whether primary or secondary, the amount of REE in these enriched deposits is still relatively low. Nature can take the processes of concentration only so far before men and women with their machines and manipulations take over. Only then can each REE be isolated, purified and put to a technological use.

In practice, the REE are chemically difficult to separate from one another. The costs of extraction and the complicated mechanical and chemical processes that have to take place before individual REE can be released therefore mean that these products are expensive to produce. REE such as dysprosium, europium and terbium can fetch hundreds, even thousands of dollars per kilogram on world markets.

China is the world's largest producer of REE. It also has the biggest reserves. Although Brazil, Russia, the USA, India and Australia have significant deposits, China currently dominates world markets. The fundamental importance of REE in modern technological developments means that those with the ore have the potential to exert political and economic influence and leverage over those who don't. For example, nearly all clean-energy products, from solar panels to electric vehicles, require considerable amounts of REE, as well as lithium and cobalt.

China is also the world's largest consumer of REE, reflecting its growing dominance in many areas of clean technology development. The electronics and engineering industries of Japan, China, Europe, Vietnam, the USA and South Korea are the biggest importers of REE (2019 figures), again hinting at the political as well as economic sensitivities surrounding our growing reliance on REE to make the modern world tick.

* * *

So far, we have touched on just a few of the uses to which REE, collectively and individually, can be put. The list is huge and continues to grow as chemists and engineers find more and more applications for these intriguing metals with their peculiar properties. Here are just a few examples in which their compounds can be found.

A major use is in the manufacture of magnets, both big and small. Magnets, especially those alloyed with neodymium, are key components in computer hard-disc drives, MRI scanners, microphones, headphones, electric motors, power generators and wind turbines. REE are used along with phosphors for use in lasers, fibre-optics, touchscreens and television screens. They occur in the catalytic converters of cars. They play a critical part in fuel cells, batteries and some superalloys. They crop up in ceramics, UV-resistant glass, X-ray imaging and liquid crystal displays (LCDs).

It is perhaps even more staggering to take just one example of an everyday object, the hybrid car, and analyse the number and types of REE that are used in its manufacture. Hybrid electric cars have cerium in their UV light-filtering windows; europium, yttrium and cerium in their sat-nav LCD screens; yttrium in the car's sensors; neodymium, praseodymium, dysprosium and terbium in their electric motors and generators; neodymium in the glass of the headlights; neodymium in all of the motor magnets that operate things like windows, windscreen wipers and on-board computers; cerium, zirconium and lanthanum in their catalytic converters; and lanthanum and cerium as well as lithium in the hybrid batteries. Quite a list. For just one car.

REE are used in a wider range of products than any other group of elements. They have become indispensable in the manufacture of a huge number of electrical, optical and power-generating devices. Their unique magnetic, luminescent and electrochemical properties have meant that many machines and gadgets can be made smaller and lighter. They can operate with less energy and greater efficiency. 'In these applications,' says the British Geological Survey, 'REE play a vital role in environmental protection, improving energy efficiency and enabling digital technology.' That is some achievement for a series of elements initially named rather unpromisingly as the 'rare earths'.

So, we see, the modern world has grown highly dependent on lithium and the REE. They not only contribute to the huge leaps we continue to see in the development of new technologies, they also play a vital role in the growth of clean-energy products, including electric cars, solar panels and wind turbines. If the world is to go 'green', the demand for lithium and the REE will continue to grow and grow.

That's the bright side. But there are clouds on the horizon. We have already alluded to the economic and political implications of the REE being rare. The fact that economic concentrations of their ores occur in only a few countries means that they can be used as a weapon in hostile trade negotiations, especially in those cases where high-tech manufacturing countries don't have REE mines of their own.

Although prices fluctuate, the predictions are that there is only one way prices will go over the long term: up. Mindful of these facts, the late Chinese leader Deng Xiaoping said, 'The Middle East has oil, and China has rare earths.' Also mindful of these facts, the rest of the world has become very active in looking for new deposits. Over recent years, exploration geologists have been busy mapping the ground for REE from Greenland to Kazakhstan, from Canada to South Africa in the hope that new deposits will help balance the political as well as the economic landscape.

Ironically, given their green credentials, the other problem associated with REE relates to their low concentrations, even in the rocks that are mined for their minerals. In order to produce commercially economic amounts, vast tracts of earth and rock have to be dug, scooped and scraped to begin the concentration process. The environmental damage can be considerable. Some ores also contain radioactive elements such as uranium and thorium. In Malaysia, these hazards posed sufficient danger to render some mining operations inadvisable.

A number of strong chemical reagents are used in the refining process, which, if not carefully controlled, can harm those who work with them. Waste gases only add to the problem, both for workers and the local environment. The final irony is that huge amounts of energy are used to extract and process the REE ores. In China, most of this energy is generated by coal-fired power stations. These produce enormous amounts of carbon dioxide, adding to the volume of greenhouse gases in the atmosphere. China is aware of the problems and its irony. As a result, it is determined to become a world leader in the production of green energy, all with the help of its home-mined, home-produced REE.

Lithium and the REE, like many of the materials we have been considering in this book, seem to live lives in which they are both saints and sinners. Whether it's oil or aluminium, iron or iridium, it is nearly always a case of the good, the bad and the ugly. Although we have acknowledged some of the downsides in our look at how we manufacture our material world, the story of stuff has largely been one in which we have celebrated humanity's cleverness as it has learned to turn rocks into things – things of every conceivable kind. However, it is now time to recognise that all of this success has come at an increasing cost, both to people and to the planet.

Chapter 11

Pollution and
the Wounded Planet

The nineteenth-century British polymath John Ruskin
was among the first to recognise and then worry about
the effects of the Industrial Revolution on the environment. Like the Romantic poets, he believed that there
was a deep, organic relationship between people and the
world in which they lived. If that world was rich in beauty
and natural wonders, so much the better for the spirit
and the soul. But if that world was ugly and disfigured,
the lives of men and women would be impoverished and
imagination crushed. Unrestrained industrialisation and
its dirty factories ruined minds as well as meadows.

In his 1860 essays, *Unto This Last*, Ruskin rails against the
social wrongs of capitalism. He anticipates many of the argu-
ments that would later be made by the Green Movement.
Industry's greed and rapacious disregard for the damage it
inflicts on the natural world distressed him:

> *All England may, if it so chooses, become one manufactur-
> ing town; and Englishmen, sacrificing themselves to the good
> of general humanity, may live diminished lives in the midst
> of noise, of darkness, and of deadly exhalation. But the world
> cannot become a factory nor a mine.*

In one of his later lectures, 'The Storm-Cloud of the Nineteenth Century', delivered in 1884, Ruskin felt that human progress would foul the world, leaving nothing but a 'blanched sun' and 'blighted grass'.

Although the word 'ecology' was not yet in common usage, the Romantics, including Wordsworth and Coleridge, were growing increasingly bothered by humanity's greedy exploitation of the Earth and its resources. Industrialisation and urbanisation were alienating men and women from nature. They felt they were witnessing a breakdown between people and the natural world. In this sense, they were ecological pioneers. They were developing, writes their biographer Jonathan Bate in *The Song of the Earth*, an 'ecological consciousness'. Nature's gifts should be respected and nurtured. They understood that we are intimately, unavoidably, part of nature's dense fabric. When men and women set themselves apart from, and above, nature, they break the organic unity that helps keep the world in balance and harmony.

Most of the world's population now live in cities. These man-made environments alienate us from nature. They are tamed landscapes of concrete, tidy parks and mown grass. In his 1969 book, *The Making of a Counter Culture*, the American historian Theodore Roszak wrote presciently that an alienated:

> ...*attitude toward the natural environment comes easily these days to a population largely born and made in the almost totally man-made world of the metropolis... The flora, fauna, landscape, and increasingly the climate of the earth lie*

practically helpless at the feet of technological man, tragically vulnerable to his arrogance. Without question, we have triumphed over them ... at least until the massive ecological consequences catch up with us.

Roszak said this more than fifty years ago, and things have only got worse.

In the natural world, there is a vast interconnectedness between things. In any complex system, change one thing and you change everything else. Feedback loops, knock-on effects and consequences, for good or ill, all feature whenever a system is disturbed, disrupted or distressed. This is how the journalist Jonathan Watts describes nature's tipping points and feedback loops as greenhouse gas emissions increase and the planet heats up:

Tipping points ... are reached one after another: methane release from permafrost; die-off of the tiny marine organisms that sequestered billions of tonnes of carbon; the desiccation of tropical forests. People have come to realise how interconnected the world's natural life-support systems are. As one falls, another is triggered... In some cases, they amplify one another. More heat means more forest fires, which dries out more trees, which burn more easily, which releases more carbon, which pushes global temperatures higher, which melts more ice, which exposes more of the Earth to sunlight, which warms the poles, which lowers the temperature gradient with the equator, which slows the ocean currents and weather systems, which results in more extreme storms and longer droughts.

Today, there is no getting away from the fact that the biggest disturbers of the Earth and its extraordinarily complex, interlocking systems are human beings as we continue to transform and transmute nature from one thing into another. Humanity has become, says the writer and environmentalist Gaia Vince, 'a geological force'. We scrape, mine and dig; we move, build and burn. *As a species we now shift many times more rock, soil and sediment each year than the world's rivers and glaciers, wind and rain combined* – to paraphrase Vince's conclusions. Impressive, but there are consequences. The land degrades. There is a loss of biodiversity. Rivers and seas become polluted. And, most profound of all, the air warms and the climate changes.

Nemonte Nenquimo is a mother and a leader. The Amazon rainforest is her home, as it has been for her people for thousands of years. Writing in *The Guardian* newspaper, she says:

> It took us thousands of years to get to know the Amazon rainforest. To understand her ways, her secrets, to learn how to survive and thrive with her... When you say the Amazon is not burning, we do not need satellite images to prove you wrong: we are choking on the smoke of fruit orchards that our ancestors planted centuries ago. When you say that you are urgently looking for climate change solutions, yet continue to build a world economy based on extraction and pollution, we know you are lying because we are the closest to the land, and the first to hear her cries.

Easier than digging deep holes in the ground, surface mining is also cheaper. Lower costs mean lower prices. This is why, whenever possible, mining companies prefer to extract ores from the surface. To get at the ore, trees might be felled, top soils stripped, surface rocks ripped and roadways built to carry the ore to where it is to be processed. The soil, trees and grass removed are euphemistically described as 'overburden'. Many open-cast mines gouge vast pits into the earth. Huge, deep holes, often hundreds and hundreds of metres across, scar the landscape.

Dust, noxious gases and chemicals pollute the air and fall as a suffocating, toxic rain on the surrounding countryside, its people, plants and animals. Enormous amounts of ore waste accumulate around the excavations. The environmentalist Annie Leonard gives a nice example. To obtain enough gold to make one gold wedding ring, 20 tons of hazardous mining waste are created in the process.

Surface mining destroys places where plants grow and animals live. It amounts to habitat destruction. Once their operations have finished, the more responsible companies attempt to replace the lost habitat – if not with what was there before, then at least with something of ecological equivalence. After miners have finished their work, they might put back the leftover rock, spoils and tailings into the holes they have dug before finally covering the waste with a layer of fertile soil. Coal tips are rounded and smoothed then sown with seeds and planted with trees. A clay pit might be filled with water and become a home for plants, birds and fish. An exhausted quarry might be allowed to weather and evolve into a haven for flowers and insects. Less responsible companies might simply abandon the landscape, leaving it despoiled, poisoned and ruined.

However, no matter how responsible the mining company, habitat recovery and renewal might not always be straightforward. Stripped of vegetation, the exposed earth and dumped topsoils are prey to wind and rain. Without the protection of trees and their roots, plants and their spongy remains, soils can easily be blown and washed away. All that is left is barren rock. Fresh topsoils can be spread over the lifeless spoils, but this adds to the costs. Only in better-regulated and well-monitored regimes might the exhausted earth be returned to anything like an ecologically healthy habitat.

But while operations are still taking place, dangers are ever-present. Mountains of removed 'overburden', rock debris and industrial 'tailings' are often built close to the mines themselves. These tips can be huge in size. And because they are composed of loose material, they are inherently unstable. To be safe, care has to be taken to make sure they are not too steep or too close to vulnerable populations. When these precautions are not taken, accidents happen.

The pit at Aberfan was one of the many coal mines sunk in the valleys of South Wales. Several colliery tips had been built along the valley sides, one of which had been constructed above a natural spring. On the morning of 21 October 1966 and after a period of heavy rain, the tip had become saturated and unstable. It began to slip. A huge slurry made its way down the hill, engulfing a local primary school and other nearby buildings; 116 children and twenty-eight adults were killed in the disaster.

On 25 January 2019, a mining dam that sat above Brumadinho, a large town in south-eastern Brazil, collapsed. It released a huge wave of waste and mud that swept over homes, businesses and residents in its path, killing more than two hundred people. The dam was known as 'an upstream tailings dam'. Waste from the iron ore mine formed a muddy lake that was held back by little more than a wall of silt and sand. The belief was that the 'lake' of semi-hardened muddy iron tailings would be solid enough to contain itself. But the mud became liquid and mobile, breached the light-weight dam and swept down the valley, destroying people, property and anything else that lay in its path.

Many minerals can be extracted only by digging holes underground. Complex networks of mine shafts and tunnels burrow deep under the surface. Water that would otherwise flood the mine has to be pumped out. Very often, these waters will be saturated with dissolved, poisonous minerals. They can't simply be sloshed into local rivers and water systems without causing severe environmental damage. Similarly, many mining techniques use water to facilitate their extraction, processing and waste-disposal activities. Wastewater from these processes can poison land and pollute rivers, killing fish and aquatic plants.

The unwanted bits of the rocks extracted from mines have to be dumped in and around the surface. Many of these waste tips contain chemicals and minerals that are also poisonous to plants and animals. The tips need careful management if they are not to leak into rivers, spread across fields and pollute the air. The same is true when mining processes involve the use of toxic minerals like mercury and chemicals, including cyanide.

Once exhausted, the subterranean tunnels that riddle the earth can collapse. On the surface above the abandoned mines this can result in subsidence and sinkholes. Cracks can creep across buildings. Water courses can suddenly disappear down holes that were not there the day before. We are, in short, a species that burrows down and buries deep as well as one that builds up and spreads out. We exploit and extract, dig and delve, dump and dispose.

It behoves governments and companies to mitigate the unavoidable harm that mining causes. Although preventive and remedial actions cost money in the short term, the long-term savings are incalculable. The health of the planet as well as human health depend on how we monitor and manage the process of taking rocks out of the ground and turning them into our material world.

* * *

The environmental hazards don't stop, of course, with simply mining the ores. There is the continued risk of pollution not just during the refining and manufacturing stages of making stuff, but when we use the stuff we make, and when we dispose of it once we've used it.

Pollution occurs when a contaminant enters the natural environment. Pollution can upset the ecological balance of land, air and water. It can render an environment dirty, unsafe and unsuitable for use. The chemical fertilisers that boost crop production can wash down into rivers and out to sea, affecting both plant and animal life. Pesticides kill the insects that the birds eat. Fewer bees buzz. Our crops don't get pollinated. The skies no longer sing to the song of the skylark and swallow. This is the 'silent spring' about which Rachel Carson wrote so prophetically in 1962.

The more extreme examples of major environmental disasters become headline news. The Union Carbide plant in Bhopal, Madhya Pradesh, central India, produced pesticides. A highly toxic chemical, methyl isocyanate gas, formed part of the manufacturing process. On the night of 2 December 1984, a devastating accident at the plant released at least 30 tons of a methyl isocyanate gas into the air. The factory was surrounded by shanty towns, and more than 500,000 people were exposed to the deadly gas. The gas hovered close to the ground, flowing its toxic way into house and home. Victims suffered burning eyes, irritated throats and nausea, leading in many cases to death. Estimates of the final death toll vary, but figures as high as 15,000 often appear in government reports. The poisonous remains and ground contamination persisted for years. Increased rates of children born with physical disabilities and learning difficulties are also blamed on the disaster.

Over the days of 25 and 26 April 1986, a catastrophic accident occurred at the Chernobyl Nuclear Power Plant in northern Ukraine, then part of the USSR. A steam explosion and graphite fire sent radioactive material high into the air, which left the immediate environment severely contaminated. But the air-borne radioactive particles were blown far and wide across Russia and much of Europe. Rivers, lakes and land became poisoned. The risks of cancer increased for those living downwind of the toxic plume.

Farm animals, including sheep grazing on contaminated grassland, had to be slaughtered. Their milk and meat could not be eaten.

The oil tanker *Exxon Valdez* ran aground on Prince William Sound's Bligh Reef in Alaska on 24 March 1989. It was carrying approximately 53 million gallons of crude oil. After the accident, around 10.8 million gallons of oil sloshed out and fouled hundreds of miles of the Alaskan Coastline with a black, sticky slick. More than 250,000 sea-birds, almost 2,800 sea otters, 300 harbour seals and 22 orca whales, along with huge numbers of herring and salmon, died in the disaster. The seas, rocks and beaches are still recovering.

* * *

Although these disasters are dramatic, immediate and devastating, they pale in comparison to the insidious, long-term, cumulative catastrophes caused by humanity's failure to respect the Earth and its bounty. Greed and ignorance slowly but ineluctably poison the land, alter the atmosphere, foul rivers and pollute oceans. In many parts of the world, the basic human right to have access to clean water for drinking has become a problem. Rains wash pollutants and contaminants into rivers, lakes and groundwaters. Wells become bitter. Springs leak only iron-rusted, sour-tasting trickles.

Worldwide, billions of tons of household and industrial waste end up in landfill. Fabrics, plastics, rotten food, bottles, cans, rubber tyres, crumbling concrete, bricks, wood, broken glass – anything and everything can end up being shoved into holes or bulldozed into mountains of rubbish. And there it all remains, and unless it is carefully sealed and managed, its toxic afterlife will ooze and leak its way into groundwaters, rivers and the sea. Many of the world's rivers, especially those that flow through great cities and industrial belts, end up as nothing more than huge sewers. Marine life is hard-pressed to survive in some of the worst-affected oceans.

* * *

To illustrate the environmental problems we have created, let's consider pollution by plastic. For decades, plastic was hailed as a hero. It was, and still is, cheap, endlessly versatile and a solution to so many problems of modern life. Chemists and engineers have created thousands of different kinds of plastic to meet thousands of particular needs.

Want a lightweight, chemically resistant car bumper? Choose plastic. Need a thin, flexible little card to prove your loyalty to a shop, that you're a licensed driver, that you can put into a bank cash machine to withdraw money? Make it out of plastic and laminate it with polyvinyl chloride acetate (PVCA). Want an extraordinarily light, flexible, strong, single-use bag to carry your shopping? Think plastic, think polyethylene. Feeling thirsty? Pack a lightweight plastic bottle in your pocket. Each year, soft drinks and other bottling companies produce around 500 billion plastic (polyethylene terephthalate, or PET) bottles a year.

However, over recent years, the world has woken up to plastic's downsides. Susan Freinkel writes that plastics have become 'both the miracle and the menace of modern life'. We produce an awful lot of it. A 2020 paper by Emily Elhacham and colleagues in the journal *Nature* reports that the world's plastics alone now weigh twice as much as the planet's marine and terrestrial animals.

Plastic's virtues are undeniable, but its vices are only just becoming apparent. In our rivers, in our oceans, in our land-fill sites, discarded PVCs slowly leak a cocktail of poisonous additives into the environment. The trouble is, plastic just isn't very good at degrading. It can hang around for years, for centuries, littering the countryside, clogging up rivers, spoiling beaches, polluting the oceans. Over 10 million tons of plastic waste enter our seas every year, with the amount continuing to rise. And there it stays.

Swirling clockwise in the northern hemisphere and anti-clockwise in the southern hemisphere, five great currents circulate around the planet's oceans. These rotating ocean rivers are known as gyres. The middle of the gyres is where the currents are slackest. And gyres, particularly their still centres, are where much of the world's discarded plastic eventually ends up.

One such polluted sea has been dubbed the Great Pacific Garbage Patch, a 'trash vortex' in the north Pacific Ocean. It measures hundreds of miles across. Much of the plastic waste caught up in the gyre has its origins in the countries of Asia, *National Geographic* has reported. Although fragments of plastic bags, bottles and discarded fishing gear might be spotted, around 40 per cent of the debris is of a smaller size and harder to see. The upper levels of these wasted waters consist of a dispersed but nevertheless dangerous mix of medium-sized, small, even microscopic particles of plastic and chemical pollutants. In the centre of the gyre, there can be as much as 100 kilograms of plastic and other waste per square kilometre.

Not one of the Earth's seven seas is plastic-free. A 2020 study by Katsiaryna Pabortsava and Richard Lampitt of the National Oceanography Centre in Southampton found more than ten times as much plastic in the Atlantic Ocean than had previously been estimated. They measured the top 200 metres of the sea and found between 12 and 21 million metric tonnes of microscopic particles of the three most common types of plastic in about 5 per cent of the ocean. Scaled up, this would suggest that swilling around in the Atlantic Ocean are at least 200 million tonnes of these common plastics.

The romance of unspoiled watery wildernesses has been lost. There are no more pristine oceans. But the poetic costs are as nothing compared to the harm that plastics cause marine life. When plants and animals die, they get eaten. They harmlessly enter the food chain. Or they biodegrade. However, when plastics enter the ocean, they merely disintegrate into smaller and smaller pieces as wind, waves and sunlight pound them up and break them down.

Whether it's plastic bags or plastic microbeads used in cosmetics and toothpastes, whether it's polychlorinated biphenyl (PCB) or bisphenol A (BPA) released when plastic polymers disintegrate, they all become a danger to the health and well-being of the planet's marine life. Microplastics have been found in rainwater in the Rocky Mountains, snow in the Arctic, wind-borne dust in the Pyrenees, and at the bottom of the world's lakes and oceans.

Plastics, particularly microplastics, have now entered the food chain. Whenever humans eat fish, cook crustaceans or swallow oysters, they too will find themselves ingesting these microplastics – those same plastics that were dumped in landfills, flushed down toilets and chucked overboard when no longer needed.

The Laysan albatrosses of the remote Pacific island of Midway have a wingspan of more than 2 metres. They can glide for miles and soar high above the waves as they ride the winds without once flapping their wings. These magnificent birds can live for thirty, forty, even fifty years or more. But they have a problem. They confuse plastic waste with food, or they accidentally scoop up the debris as they skim the ocean's surface with their beaks, hunting at speed for fish and squid. And the indigestible plastics end up in their stomachs, where they slowly kill them. Even worse, when adults are raising chicks, they regurgitate the plastics they have swallowed and feed them to their young. Over recent years, the death rate of albatross chicks has risen alarmingly. The plastic fills and blocks their stomachs. It cuts the lining of their guts. And they die.

Abandoned fishing nets are a particular problem. It is estimated that over half a million tonnes of fishing gear is abandoned, dumped or lost each year. The nets are made of nylon and plastic and they ensnare thousands of dolphins, turtles and seals annually. Those creatures unlucky enough to be caught up in these 'ghost nets' suffer a slow and painful death.

In March 2019, a young whale washed up on the shores of the Philippines. Marine biologists said that it had died of 'gastric shock'. In its stomach they found 40 kilograms of plastic debris, including sixteen plastic rice sacks. Although this Cuvier's beaked whale had a particularly high content of plastics in its gut, it was by no means the first cetacean that had ended up dead on the island's beaches after ingesting the polluting polymers. Worldwide, thousands of sea mammals die each year, their stomachs full of bits of plastic. However, the problem is particularly acute in South-East Asia. In 2018, a report by the Ocean Conservancy stated that China, Indonesia, the Philippines, Thailand and Vietnam dumped more plastic in the oceans than the rest of the world combined.

Our addiction to plastic is proving a hard habit to quit. We might give up using single-use plastic bags and resist buying food wrapped in clingfilm, but it is so much harder to avoid buying one or other of the thousands of other things in which plastic plays an integral part. TV remote controls, electric cars, computer laptops, triple-glazed window frames, polyester fabrics, running shoes, toothbrushes, combs, wheelie bins. The Age of Plastic feels all-encompassing. Leading an oil-based plastic-free life is a challenge beyond most of us. A small glimmer of hope lies in the discovery of some super-enzymes and bacteria that eat and degrade plastic. But pollution by plastics is so widespread that an enormous challenge remains.

However, whether humanity as a whole is really capable of walking with a lighter tread across the planet and its resources is not certain. The philosopher John Gray certainly doubts it. In his book *Straw Dogs*, he sees little prospect of our species, with our animal instincts, holding back our appetites:

> *The destruction of the natural world is not the result of global capitalism, industrialisation, 'Western civilisation' or any flaw in human institutions. It is a consequence of the evolutionary success of an exceptionally rapacious primate. Throughout all of history and prehistory, human advance has coincided with ecological devastation.*

Let's hope he's wrong.

Chapter 12

Coal, Oil and Climate Change

Throughout the pages of this book, we have noted the key part that fire and heat play in converting rocks into minerals, and minerals into stuff. The amount of energy needed to make glass, concrete, aluminium, bricks, iron and steel is enormous. Much of this heat and energy comes from burning fossil fuels. Eighty-four per cent of the energy consumed worldwide in 2019 came from burning coal, oil and gas. About a third of fossil fuels are burnt to produce electricity.

We have already looked at oil, its formation and extraction, when we considered how plastics are made. However, only a small fraction of crude oil ends up as a plastic. Most petroleum and its products are burnt to fuel cars and lorries, ships and planes, power stations and the fires that heat kilns and blast furnaces. They fire the boilers that keep our offices and houses warm. But as well as releasing energy, burning hydrocarbons in air also produces carbon dioxide, a greenhouse gas.

Cars, vans, lorries and buses, especially those with diesel engines, also pump out nitrogen oxides and tiny 'particulates' of black carbon. These exhaust emissions are bad for our lungs, bad for our health, and bad for the climate. They contribute to the polluted fogs and smogs that descend on the world's big cities whenever the air is still.

The other fuel that we continue to burn in massive quantities is coal. It has played, and continues to play, a critical role in many of the world's major manufacturing industries. If we are to understand climate change and global warming, we need to look at this most interesting of rocks in a little more detail.

As with oil, small amounts of coal and its by-products can also be turned into material stuff. Coal tar is used to manufacture creosote oil, naphthalene, phenol and benzene. When coal is heated to make coke, one of the by-products is ammonia, which can be used to make ammonia salts, nitric acid and agricultural fertilisers. A huge variety of other products also involve coal or coal by-products in their manufacture, including soap, aspirins, dyes, plastics and carbon fibres.

You can even make jewellery out of coal. Jet is a form of lignite that has been subjected to high pressures deep beneath the earth. The famous ornamental black jet of Whitby on the North Yorkshire coast is Jurassic in age, approximately 182 million years old. These jets ultimately derive from the fossilised, lithified remains of an early relative of the monkey puzzle tree, *Araucaria*. When carved, the shiny black jewellery makes exquisitely beautiful pendants, brooches and rings. Its smooth, shiny, deep blackness is the origin of the expression 'jet black'. When Prince Albert died in 1861, Queen Victoria chose to wear Whitby Jet jewellery in remembrance of him. With royal support, it soon became the fashion to accessorise one's mourning dress with jewellery made from the black gemstone.

However, the main thing we do with coal is burn it,

releasing energy as heat in the process. It is a combustible rock. We might divide coals into two broad types. Steam coal, also known as thermal coal, is mainly used in power generation. Most of it is burnt in power stations, where it boils water into steam, the steam drives the turbines, and the spinning turbines generate electricity. The other type of coal is described as coking coal, or metallurgical coal. Converted into coke, it is mainly burnt to fire the blast furnaces of the iron and steel industry.

More than half of coal is made of the element carbon. Burning carbon in air, a fifth of which is oxygen, produces the gas carbon dioxide, whose Jekyll and Hyde character we shall examine shortly. But first, having flirted with coal's various uses, we need to look at its origins and extraction.

* * *

Coal, like a number of other useful rock types, requires biology and geology to work together before it can form. Banded ironstones for steel, limestones for cement, and oils for plastics all depend on plants and animals being caught up in mighty geological processes for their formation. These are reminders of the complex relationship between life on our planet and its rocky history.

Coal is the lithified residue of plant matter that has been preserved and altered by heat and pressure. Deposits are known from every geological age since Devonian times around 390 million years ago, when plants made their first appearance on land. Up until then, the continents had been largely rocky and lifeless, possibly save for a few pioneer mosses, bacteria and other small life forms scuttling along the margins. But after the Devonian, plants turned the rainy lands from brown to green. Coal formation became possible.

The large coalfields in Europe, Britain and North America were deposited in the Carboniferous period, beginning around 310 million years ago. Permian coal is found in Antarctica, Australia and India. Much younger Tertiary coals are found in Spitzbergen, western USA, Japan, India, Germany and Russia. Coal is found on every continent on the planet.

As an example, let's look at the geological history of the Carboniferous coal deposits that stretch west to east, from Britain, across northern France and Belgian, to northern Germany and over to Poland. And once more, we have to consider plate tectonics and the planet's ever-shifting geography.

Around 310 million years ago, the British Isles and most of Europe found themselves straddling the boundary between two tectonic plates, Laurasia (present-day Europe, Asia and North America) and Gondwana (present-day Africa, the Arabian peninsula, South America, Antarctica, Australia and India). These two plates were in the process of slowly colliding to form a new supercontinent, Pangea.

Britain and Europe were situated along the south-eastern edge of the Laurasian plate. Because this was an area that sat along the boundary between the two converging crustal slabs, it got caught up in the pulls and pushes of the moving blocks. The crust above the subduction zone began to sink as the two plates shunted into each other. This created a huge sinking sedimentary basin into which seas poured. On the continental side of the convergence zone, the less dense continental rocks were being pushed high into a mountain chain. It was the weathering of this mountain chain that supplied the sediments that were filling the basin. However, as the tectonic forces continued to push and plunge the two plates together, the basin was sinking as fast as it was being filled. This resulted in the accumulation of thousands and thousands of feet of sediments over millions of years.

At first the basin seas were wide and open, rich in corals and sea life. The marine deposits during this time were destined to become the Carboniferous Limestones.

These shallowing and silting seas were followed by deltas and rivers that brought sands and grits down from the nearby mountains. These deposits we now see as the tough, dour Millstone Grit country of the Pennines.

The final stage of these Carboniferous days was a time of swamps and warm, shallow seas and lagoons. The key point to remember, though, is that these deltas, swamps and seas were tropical. Some 300 or so million years ago, the eastern parts of North America, the British Isles and northern Europe *lay across the equator* on the southern margins of the continent of Laurasia.

The wet and warmth, the rivers and swamps, were the perfect place for plants to grow big and in abundance. The rich vegetation was very different from what we see in today's equatorial forests. Lush swamp forests of giant tree ferns and horsetails stretched as far as the eye could see. There were fish in the rivers and lakes. Frighteningly large newt-like amphibians waded through the muddy waters. Because of the exceptionally high levels of oxygen in the atmosphere during Carboniferous times, super-sized insects buzzed above. Fossil dragonflies measuring 75 centimetres across have been found preserved in the muds, silts and coalified peats.

From time to time, the sea would flood these slowly sinking forested swamps, killing the vegetation and depositing thin beds of mudstone, called 'marine bands', rich in fossils. And then the lakes and deltas, swamps and fern forests would return once more. This cycle of plant-rich swamps followed by warm, shallow silty seas would be repeated many times, resulting in cycles of sedimentation: peats (destined to turn into seams of coal), silts, sandstones, limey shales and back to compacting peat. Today we see these sedimentary cycles as the beds of the Carboniferous Coal Measures.

The tropical, plant-rich swamplands provided the perfect environment for the eventual formation of coal. The first stage of coal formation requires large amounts of dead plant debris to accumulate year on year under conditions that preserve it. Brackish, swampy waters are ideal. Here, there is little or no oxygen. These stagnant, anaerobic conditions prevent the organic matter from oxidising. This also means that nothing else can live there, so there are no organisms to eat the dead vegetation. Moreover, and critically, we have to remind ourselves that these swampy basins were sinking at roughly the same rate as the silts and peats were accumulating. This allowed thousands of feet of sands, silts and coal seam sequences to build up.

The next stage requires the thick layers of decaying vegetation to be buried deep enough for the weight and pressure above to drive out water and other liquids. As the dead plant matter gets buried deeper and deeper, a sequence of 'coalification' begins. It starts with the formation of peat, then lignite (brown coal), followed by bituminous or soft coal, and finally the highest-quality coal, known as anthracite. With each stage, the amount of carbon content increases.

Although peat and brown coals can appear under relatively shallow burial conditions, higher-quality coals and anthracite need much higher levels of heat and pressure before they can form. These can be achieved only in more dynamic geological environments. The most active regions are found where two tectonic plates converge, plunging the basin sediments that accumulate between the two plates deep down into sinking crustal blocks and subduction zones. If these sediments contain thick layers of peat, they too will be subjected to great pressure and heat. This is exactly what was happening around the equator 300 million years ago as the plates of Laurasia and Gondwana collided.

Over the next few million years, the buried sediments of these Carboniferous times cemented and hardened. They turned into rocks, including limestones, sandstones, shales, grits and, in the case of the swampy peats, coal.

In the following 250 million years, the continental plate of Laurasia, of which North America, the British Isles and northern Europe were still a part, continued drifting north. Then around 65 million years ago, the North American and South American blocks of the fragmenting plates of Pangea and Laurasia began to split, drifting off westwards. Diverging mantle convection currents running north to south beneath Pangea and Laurasia created a huge rift into which the seas flowed. As the two plates pulled apart, the seas became wider and deeper, creating a new ocean, the Atlantic. Today, the northern European part of the Eurasian plate lies roughly 50 to 60 degrees north, a long way from the equator. Lying several thousand miles west of Europe, and still drifting away at a rate of about three or four centimetres a year, lies the North American plate with its own beds of Carboniferous coal.

* * *

Plants thrive and grow by using light and energy from the sun to turn atmospheric carbon dioxide into leaves and wood. This is the process of photosynthesis. Coal, and the energy locked within it, really is not much more than trapped sunlight. When we burn the carbon in coal, we are releasing that stored sunlight, that ancient energy. The knowledge of how to turn coal back into heat and light has a very long history.

Being a rock, like any other rock, coal beds occasionally crop out at the surface. Prehistoric men and women were the first to realise that this black, flaky rock could be burnt. It was easily mined, and for many centuries it enjoyed regular, but modest, use as a fuel. By the time of the ancient Greeks, Romans and early Chinese dynasties, coal was also being used to fire metal-smelting furnaces, but it had yet to replace wood and charcoal on any major scale. The Romans also began to burn coal to heat water for their public baths and to warm air to heat the floors of their cold northern villas.

From the twelfth and thirteenth centuries onwards, worldwide coal production and use slowly began to increase. It was mainly being burnt in furnaces to smelt metals. Small adit mines opened up wherever coal seams broke the surface. For example, the Abbot of Dunfermline was granted a charter in 1291, allowing his monks to dig for coal in Pittencrieff, Fife. Coal's use for domestic heating, cooking and firing pots also began to gain in popularity, not just in Europe but in North America and Asia, too. In sixteenth-century Britain, supplies of wood began to falter. Hardwoods, such as oak, were needed to build houses and ships. Burning coal therefore began to replace burning wood.

As the need and uses for coal grew, mining became an important and profitable activity in its own right. In Great Britain, deep-shaft mining didn't become widely practised until the late eighteenth-century. Mining deep underground was a hazardous business. Mines could flood. They could fill with poisonous gases. They might collapse. On the plus side, though, the engineering, economic and safety needs of the coal industry were spurring invention.

Steam engines were invented to pump water to the surface and ventilate the shafts. Newcomen's famous steam engine made its first practical appearance in 1712. In 1815, Humphry Davy invented his safety lamp to help prevent gas explosions. Timber supports propped up roofs. By the 1700s, Britain was producing up to 3 million tons of coal a year.

Perhaps one of the biggest boosts to coal mining, which helped bring about the Industrial Revolution, was the development of the modern steam engine, invented by a Scottish engineer, James Watt. The steam to run his engines was produced by burning coal to boil water. By the late 1770s, Watt had made a powerful, reliable engine that could pump water and drive machinery. Other inventors would go on to develop and refine the engine, but the basic principles were those originally pioneered by the Scotsman. By the nineteenth century, steam engines were powering locomotives that ran on rails, propelling ships across the oceans, and running the machines that were speeding up the rate at which factories could operate and churn out their wares at ever-cheaper prices.

Throughout the late eighteenth and nineteenth centuries, coal was also being used in increasing amounts to make coke to fire the furnaces of the iron and steel industry, the kilns of the pottery factories, and the heat needed to run the growing chemical, glass and cement plants. The Industrial Revolution was well underway and, in Great Britain, Carboniferous coal was fuelling its fury.

By 1900, coal production worldwide was approaching a billion tons a year. In 2019, the world produced almost 8 billion tons, according to the *International Journal of Coal Geology*. The total amount of coal that has been burnt since the Industrial Revolution adds up to hundreds of billions of tons.

Today, China is both the world's leading producer and consumer of coal. But even before coal is burnt, its recovery pollutes the atmosphere. Methane, the bane of any miner's life, is not only highly flammable but is much more potent than carbon dioxide as a greenhouse gas, albeit shorter-lived. Methane gas tends to saturate most deposits of coal. Mining releases the trapped gas. A 2019 report by the International Energy Agency found that whenever coal is mined, huge amounts of methane escapes into the atmosphere, adding significantly to global heating.

* * *

The initial worries about burning coal concerned the soot-iness of its smoke. The soot from factories and chimney smoke from domestic fires could turn things smudgy and black. Clothes hung out to dry, stone-clad public buildings, cars left in factory car parks could all become dirty and black with settling soot.

However, the biggest worries were about smog, a lethal combination that occurs when smoke from coal mixes with fog. Coal smoke is a combination of soot, particulates and harmful gases such as sulphur dioxide. On still, damp days when mists formed, smoke particles would act as a nucleus for increasing the density of the tiny fog droplets. This reaction also meant the fog became dirty as well as thick. In extreme cases, visibility would drop to only a yard or two. These were the notorious 'pea-soupers' of the 1950s.

As a child growing up at this time in the suburbs of industrial Manchester, I well remember smogs so dense that you had difficulty finding your way home from school. Fog-beaded clothes would become filthy. If you blew your nose, the handkerchief would turn black with the foggy soot that you had been breathing in. Being a child who enjoyed extreme weather events, I rather stupidly thought smog to be quite an exciting phenomenon. The middle of the day turned to night, everything came to a stop, there was no bird song, and the world became eerily silent as the muffling, dirty air dampened all sound.

The London smogs of the 1950s were particularly bad, most notably the 'Great Smog' of December 1952. Cars couldn't run, trains had to stop, and people suffered life-threatening respiratory problems. Thousands died. Air pollution had become a major problem. The solution was the Clean Air Act of 1956. Only approved smokeless fuels such as coke or gas could be burnt in smoke-controlled areas. In most cases, this meant that coal could no longer be used as a domestic fuel. As a result, air quality, particularly in the big industrial cities, began to improve.

Although the weight and volume of carbon dioxide produced when coal is burnt depends on its carbon content, roughly speaking, burning 1 ton of coal creates just under 3 tons of carbon dioxide. The amount of carbon dioxide produced weighs more than the amount of the original fuel itself because during complete combustion each carbon atom in the coal combines with two oxygen atoms from the air to make carbon dioxide (CO_2). The atomic weight of oxygen atoms (16) is greater than those of carbon (12), and there are two oxygen atoms for every one of carbon in the gas. Hence, the increased weight differences before and after burning. Similar chemical reactions occur when we burn oil and natural gases, whether it's in domestic gas boilers, cars, ships, planes, trains, power stations, cement factories, glass works or blast furnaces.

All of this is to say that since the Industrial Revolution we have seen dramatic increases in the amount of carbon dioxide being pumped into the Earth's atmosphere, all as a result of human activity. In 1900, it is estimated that the world emitted 2 billion tons of carbon dioxide into the air. By 2018 this had risen to 36 billion tons. Over the thirty-year period between 1990 and 2020, the world emitted as much carbon dioxide as it did in all of human history up to that point.

The oceans absorb a great deal of the atmospheric carbon dioxide. This results in the seas becoming gradually more acidic, creating major problems for many marine species. Some atmospheric carbon dioxide is used by plants as they photosynthesise. However, most of the gas remains and accumulates in the atmosphere.

For most of recent human history, at least until around 1750, levels of carbon dioxide in the atmosphere remained fairly stable at around 270 to 285 parts per million (ppm). However, since we began burning fossils fuels to power our modern world, concentrations have been increasing rapidly. By 1958, levels had reached 315 ppm.

Since the beginning of the Industrial Revolution, we have released 2.2 trillion metric tonnes of carbon dioxide into the atmosphere. As capitalism's relentless logic leads to fewer and bigger industrial conglomerates, Richard Heede of the Climate Accountability Institute has found that just twenty fossil fuel companies, more than half of them state-owned, have produced 35 per cent of the carbon dioxide and methane released by human activity since 1965.

Current levels of carbon dioxide in the air have now

surpassed 400 parts per million and are well on their way to reach 415 ppm and beyond, higher than at any time during the past few million years. That is a rise of more than 40 per cent in a mere couple of centuries. This rise, coupled with increases in the emissions of other greenhouse gases including methane and nitrous oxide, has led to a rise in average global temperatures.

Climate scientists have found that since pre-industrial times, average global temperatures have risen by approximately 1.2°C and unless we drastically reduce greenhouse gas emissions, this rise is likely to continue. The decade beginning 2010 was the warmest on record, culminating in 2019 as the hottest year since 1850 when global temperatures were first measured. Records for highest average temperatures continue to be broken year after year, locally and globally. In the first half of 2020, temperatures in the Siberian Arctic were up to 5°C above average. On 20 June of that year, a monitoring station in Verkhoyansk, 670 N, registered a record high of 38°C.

Glass in a greenhouse allows heat radiation from the sun to enter and warm the plants, pots and soil inside. The warm pots, plants and soil in their turn radiate energy back, but at wavelengths longer than those that entered the greenhouse. These longer-wavelength radiations are unable to penetrate glass. Therefore, much of this longer-wavelength heat radiation remains trapped in the greenhouse, raising the temperature of the air within. That's why we build greenhouses. Certain gases in the Earth's atmosphere behave in a similar way to the glass in a greenhouse, letting heat in but not letting it out, hence they are often called 'greenhouse' gases.

The main greenhouse gases in the Earth's atmosphere are water vapour, carbon dioxide, methane, nitrous oxide and ozone. The Earth receives energy from the sun in the form of ultraviolet, visible and near-infrared radiation. About a quarter of the sun's energy is reflected directly back into space by clouds and the atmosphere itself. About a fifth is absorbed by the clouds and air. The remaining 50-odd per cent of this solar energy has wavelengths that are relatively short, which allows them to pass through the atmosphere reasonably freely. The sun's heat and light rays eventually hit the Earth's surface, where they are absorbed, heating up the land and sea in the process.

Because land and sea surfaces are colder than the incoming sun's rays, when they re-radiate the heat and light, they do so at wavelengths longer than those that they absorbed from the sun in the first place. Most of this heat radiation is reflected back into the air where, because of its slightly longer wavelengths, it gets absorbed by the atmosphere's various gases, especially those of carbon dioxide, methane and nitrous oxide – the greenhouse gases. The re-radiated heat energy therefore gets trapped. This raises the temperature of the air and the atmosphere warms.

Without any greenhouse gases, the average temperature of the Earth's atmosphere would be well below zero. However, because the atmosphere is composed of nitrogen, oxygen and a small percentage of naturally occurring greenhouse gases, it does absorb some of this reflected radiation. This has meant that the historical global average for the past few thousand years has hovered at around 15°C. Until now.

Recent dramatic increases in the concentration of the energy-absorbing greenhouse gases in the atmosphere therefore account for the current unprecedented rise in global average temperatures. More heat is being trapped. This phenomenon has popularly been known as *global warming*. However, with alarm growing about the catastrophic consequences of pumping ever more carbon dioxide and methane into the atmosphere, the language has intensified. Activists and many scientists now prefer to talk of *global heating*.

Higher average temperatures heat up oceans, which, because water expands as it heats (at least once it's above 4°C), leads to rising sea levels. The heat held by the world's oceans reached a record level in 2019. Hotter seas release more moisture into the atmosphere. Hotter, wetter atmospheres hold more energy. This results in more turbulent, disturbed and extreme weather. Which leads to more severe storms, heavier rains and increased flooding in some parts of the world, while other regions become hotter, drier and subject to prolonged droughts.

Hotter air and warmer oceans also begin to melt glaciers and the polar ice caps, leading to even higher rises in sea level, threatening to drown coastal town, cities and agricultural lands.

All of this we call *climate change*.

* * *

The causal links between burning fossil fuels, increased levels of carbon dioxide, global warming and climate change are now indisputable. What we do about the threat is another matter. The political will to reduce carbon emissions and the many other pollutants with which we poison the world is still patchy and inconsistent.

A Scottish lawyer, the late Polly Higgins, was tireless in her campaign to get the crime of 'ecocide' recognised and incorporated into international law. For too long, powerful people and politicians, international companies and states, many of whom are also climate-change deniers, have ridden roughshod over environmental concerns in the pursuit of profit and power. In effect, they commit crimes against both nature and humanity. If the Earth is to be protected, they need to be held accountable for their actions and their consequences. If we fail, then some, such as the climate science journalist David Wallace-Wells, believe the world for humans and thousands of other species will become uninhabitable. Things are already bad, he says, and probably 'worse, much worse, than you think'.

We need energy. But not from fossil fuels. Renewable and 'green' energy generation must be the future, and our salvation. The technology behind these has come on leaps and bounds. But there is an economic, political and practical challenge. Today, the world uses the equivalent of 12,000 million metric tonnes of oil equivalent a year. The environmental science and political commentator Roger Pielke calculates that if we are to wean ourselves off our dependency on fossil fuels and replace this energy equivalent with, say, wind turbines, we would need to build up to 1,500 new turbines *every day* over the next 11,000 days if we are to reach 'carbon zero' by 2050 (in a 2019 *Forbes* article). That's a mighty challenge. This target can be reduced if we also adopt other policies. 'Greener' cars, ships and planes; carbon-capture technologies; the imposition of higher taxes on all carbon emissions; less travel – these can all help. To stabilise temperatures, we need to be carbon net zero. This means that the amount of carbon dioxide emitted must equal the amount of carbon dioxide removed from the atmosphere. More problematic, though, is the fact that we haven't yet worked out how to make iron, steel and cement without the help of huge amounts of fossil fuels.

Nevertheless, wind, solar and tidal power are certainly capable of generating a good deal more of our energy needs, but again, their widespread adoption requires politicians to be bold and brave. We can only hope.

Chapter 13

The Anthropocene

Throughout these pages, we've mentioned many major rock types and geological periods. The Earth is about 4.6 billion years old. To help find their temporal bearings in this vast timescape, geologists have divided these billions of years into different eons, eras, periods and epochs. It is the names of the geological periods and the type of rocks they contain with which most of us will be familiar – Cambrian, Jurassic, Cretaceous and all the rest.

As many of the nineteenth century's founding figures of geology were British, some of the names they gave to these periods were based on those parts of the country where they felt the rocks were best typified. Cambria is the Latinised version of *Cymru*, the Welsh name for Wales, hence Cambrian in honour of the splendid rock sequences of that age found in that country. The Ordovician and Silurian were named after ancient Celtic tribes who lived in North Wales and nearby Shropshire. The Devonian needs no explanation.

The periods themselves are further subdivided into 'epochs' and 'ages' right up until the present day. Geologists define our time as the Holocene epoch. It began 11,650 years ago, towards the end of the last Ice Age and the beginning of the current interglacial. In turn, along with the Pleistocene, the Holocene is one of the two subdivisions of the Quaternary period, which began about 2.2 million years ago, the beginning of the last Ice Age.

The shift from one period to another is usually marked by a significant change or sudden break in the type of rocks being laid down, say from marine sediments to volcanic lavas, or from desert sandstones to deltaic river grits.

Geologists also use radiometric dating techniques to age rocks. All rocks and minerals contain tiny amounts of a number of radioactive elements. Radioactive elements and their isotopes are unstable. Over time, they break down spontaneously into more stable, lighter atoms, a process known as radioactive decay. Radioactive decay occurs at a constant rate, specific to each radioactive isotope. This is known as the isotope's 'half-life', the amount of time it takes for half of the radioactive isotope to break down into more stable atoms. By measuring how much of a particular isotope and the products of its breakdown are present, it is possible to estimate the age of a rock. Dating using uranium's breakdown into lead is often used for rocks hundreds of million years old. In contrast, radiocarbon dating, sometimes known as carbon-14 dating, is preferred for much younger deposits, including those of very recent times as well as archaeological finds.

* * *

However, by far the oldest – and still widely used – method for deciding the order and age of a bed of rocks is to look for fossils. Major geological changes are indicated whenever there is a rapid increase or decrease in the types of fossils being preserved.

Fossils, being the petrified remains of plants and animals, tell us a lot about the state of the world at the time they were living. Major changes in the planet's geography, atmosphere and climate correlate with changes in the fortunes of different plant and animal species, both on land and in the sea. Sudden changes in the fossil record are therefore frequently used to mark the end of one geological age, epoch, period or era and the beginning of another.

A huge increase in the numbers and varieties of fossils found in a rock is one indicator of when a new period might begin. In contrast, *mass extinctions* of vast numbers of species are used as another geological marker, usually defining when a major period is coming to an end. Mass extinctions might be caused by huge volcanic eruptions pumping billions of tons of carbon dioxide, methane and sulphur dioxide into the atmosphere; or by huge releases of seabed methane gases; or by particular configurations of tectonic plates and the position and size of continental land masses; or even by catastrophic asteroid strikes. Each of these events can radically and suddenly alter the Earth's climate, the depth and size of the oceans, and the geography of the land.

When most species struggle to survive and adapt to these changes, we witness a mass extinction, the biggest of which happened 252 million years ago. The boundary marking the end of the Permian and the beginning of the Triassic period is known as the Great Dying. More than 90 per cent of all species worldwide were wiped out in only 60,000 years, which in geological terms is a very short space of time.

Let's take two other well-known examples of fossil numbers rising and then falling.

The first has been called the 'Cambrian Explosion'. In the rocks of late Pre-Cambrian times, most organisms were simple, composed of individual cells occasionally organised into colonies. This was a time of bacteria, algae, sponges, flagellates and jelly-like creatures. Then, about 542 million years ago, there was a sudden and dramatic increase in the number, range, complexity and type of species swimming in the seas and scuttling across the bottom of the oceans.

Among the complex, multicellular animals that evolved during this period were the chordates, animals with a dorsal nerve cord; hard-shelled brachiopods, which resembled clams; molluscs and corals; and trilobites and other arthropods – the ancestors of spiders, insects and crustaceans. In just a few million years, the rate of species diversification accelerated exponentially and introduced the forebears of most of the life forms we see today. This fundamental change in the size and variety of species in the fossil record marked the beginning of the Cambrian period.

The second example occurred at the end of the Cretaceous period. This also marked the final days of the Age of the Dinosaurs. Sixty-six million years ago, there was a mass extinction event. Up to three-quarters of all plant and animal species died out. Not only was this the end of the Cretaceous period, it was the end of the Mesozoic era and the beginning of the Cenozoic era, the rise of mammals and flowering plants, and the geological era in which we find ourselves today. The Holocene is the latest epoch of this era.

In the geological record that marks the end of the Cretaceous, we find a worldwide thin layer of a particular sedimentary clay. It is relatively rich in the rare earth metal iridium. Although the metal is rare in the crust, it does occur in higher concentrations in some asteroids. This got scientists thinking. Maybe the Earth was hit by a particularly large asteroid, which would have had a devastating impact on the life of the planet.

The idea was originally proposed by Luis Alvarez and his team in 1980. A decade or so later, the remains of a giant impact crater, some 180 kilometres wide, were found on Mexico's Yucatan Peninsula. It was named the Chicxulub crater. To have made such a large hole, estimates vary, but the asteroid is thought to have been between 10 and 16 kilometres across. On impact it would have vaporised and vitrified billions of tons of crust, as well as blowing apart its own iridium-rich rocky fabric. The dust, gases and aerosols released would have been blasted high into the atmosphere, and then spread worldwide, blocking out sunlight for years after.

This was also a time of massive lava floods and volcanic activity in the Deccan region of India. These eruptive events can only have added to the levels of toxic atmospheric gases and dust that were stopping much of the sun's energy from reaching the Earth's surface. The result of all these global disruptions was a loss of heat from the sun reaching the surface, a reduction in the level of light, and an upset in the ability of plants to photosynthesise. The Earth's climate became more wintry. All of which led to a sudden, geologically speaking, massive loss of plant and animal life, including the death of the non-avian dinosaurs.

As the dust slowly settled, it formed a worldwide layer of iridium-rich sedimentary clays. Taken together, this mass extinction event and the thin layer of iridium sedimentary rocks mark the end of the Cretaceous period. It also gave an evolutionary opportunity to a previously unspectacular class of animals, the mammals. In the following period, the Paleogene, mammals quickly diversified and grew in number, size and variety, leading one day – a mere 200,000 years ago – to the emergence of a new mammalian species of ape, Homo sapiens. We made our first appearance on the plains of north-east Africa. However, even though the dinosaurs were wiped out in this mass extinction event, it has to be remembered that they had been an extremely successful species. This branch of the reptile animal class had been around for nearly 200 million years. There's even one dinosaur lineage that managed to survive the Cretaceous apocalypse – the avian dinosaurs, the ancestors of all modern birds.

We might take a couple of things away from all of this. Life forms can leave their mark as fossils in the geological record. And even the most successful species can suffer sudden extinction when conditions radically change. Complacency might not have killed the dinosaurs, but we might remind ourselves that past success is no guarantee of future survival.

* * *

In less than 200,000 years, our species, Homo sapiens, has grown in number from a relatively few families of hunter-gatherers scattered around the Horn of Africa to almost 8 billion people who populate the world today. In my life-time, since the late 1940s, there has been a *three-fold* increase in the world's population. Current projections estimate that there might be as many as 10 billion people on the planet by the year 2060, although there are those who think that declining birth rates in the world's more advanced econo-mies could put a brake on this human explosion.

As a species, human beings have also spread worldwide. We have learned to live in almost every environment, from tropical jungles to temperate savannahs, from deserts to the icy Arctic, from river deltas to high mountain plateaux. We have proved to be an extraordinarily adaptive, resourceful and inventive species, and we now dominate the planet. And we are certainly leaving our mark.

We chop down forests and farm on an industrial scale. We mine for minerals and quarry rocks, moving billions of tons of the Earth's crust from one place to another. In just the past forty years we have tripled our use of natural resources.

We make millions of tons of new stuff every year. When we have finished using the stuff, we throw it away, bury it or dump it out at sea. We dig coal out of the ground and pump oil from the depths, then burn them. So many of our activities produce compounds and chemicals that pollute the air and foul our rivers, contaminate the land and poison the sea.

Such is the scale of our making and growing, digging and burning, moving and dumping, poisoning and polluting that the surface of the planet is being changed at a rate that is beginning to outstrip nature itself. This has led geologists to the idea that the Earth is being changed and scarred so profoundly by our actions that a record is being left in the planet's geological history, one that is man-made. Humanity is building the 'geology of the present'.

Many now believe that the scale of these changes is so apparent and widespread that a new geological boundary is being marked and laid down, one that highlights the impact that one species is making on the composition and charac-ter of the Earth's surface. We are that species. Homo sapiens, human beings, men and women, the Greek for whom is *anthropos*. This gives us many words in which we wish to convey the idea that the subject 'pertains to man or human beings', words such as anthropology, anthropomorphic and anthropogenic. Geologists shocked by humanity's dramatic effect on the environment in 'recent times' believe that we have now entered a new geological epoch. So they have named it the *Anthropocene*.

* * *

It is true that ever since our evolutionary emergence, we have interacted with and exploited the environment. However, for a long time, these changes were local and small in scale. They might, or might not, have left traces in the Earth's record. Flints were mined and knapped. Trees were chopped and burnt for fuel. Clay was dug to make pots. But these were mere environmental pinpricks count-ing for little in the vastness of geological time.

In this sense, for tens of thousands of years we were not so very different from the many other species who busy themselves above and beneath the ground. On the sea floor and over the land, worms have been burrowing for eons. Beavers cut down trees, build dams and trap sediments. Meerkats dig complex underground networks beneath the savannah's dry sands. In a million years' time, there might be a fossil record of any one of these thousands of events, but most will be lost and worn away, leaving no trace.

However, as human beings spread across the Earth and their numbers slowly grew, their assault on the planet and its life gradually increased. Even in Palaeolithic times, some animal species, especially large mammals, were being hunted to extinction. There are no more herds of beefy aurochs and it is several thousand years since the last mammoths roamed Siberia.

Ten thousand years or so ago, at the end of the last Ice Age and the beginning of the Holocene, men and women began to settle and farm. More and more trees were cut for timber and fuel. Forests were cleared for agriculture. Minerals began to be mined and metals smelted. The first cities appeared, built of stone, timber and bricks of clay. We were certainly having an impact on the environment, but again the scale was limited and small. And that's roughly how things stayed until recently.

By the eighteenth and nineteenth centuries, though, a few scientists began to wonder whether human beings were, in fact, beginning to modify the landscape to such an extent that the world was literally being changed under their feet. In 1778, Georges-Louis Leclerc, Compte de

Buffon, was talking of the 'seventh and last epoch' in which he felt 'the entire face of the Earth today bears the imprint of human power'. In 1854, Thomas Jenkyn, theologian and geologist, suggested that our impact on the landscape might well warrant calling the present age the Anthropozoic. Their suggestions and fears were noted but not taken too seriously.

However, a century later, the extent and depth of humanity's impact on the Earth's natural systems was becoming more apparent. It was getting harder to ignore. The pace of change, the growth in the world's population and the increasing power of science and technology to transform nature were now of a size and scale that could no longer be ignored.

Paul Crutzen is a Dutch, Nobel-prize winning atmospheric chemist who, along with his colleagues, has conducted research into the depletion of the ozone layer. In 2000, he was attending an international conference of Earth system scientists in Mexico. The scientific community was becoming aware that men and women were changing not only the atmosphere but also the world's geosphere and biosphere too, and there was a lot of talk about human-driven change throughout the Holocene right up to the present day. Feeling ever-more frustrated, Crutzen eventually interrupted proceedings, saying that maybe we were no longer living in the Holocene. He paused, thought, and then in a moment of inspiration said that perhaps we should say that we are now living in a new age, the Anthropocene.

Crutzen soon discovered that the term 'Anthropocene' had actually been coined before, by an American ecologist,

Eugene F. Stoermer, back in the 1980s. However, the term hadn't been picked up and popularised. The Dutchman contacted Stoermer and together they wrote a brief, but key, paper introducing the concept, simply titling their piece 'The Anthropocene'. The paper was published in the *Global Change Newsletter* in the year 2000.

Two years later, Crutzen published a second short paper. It proved to be a game-changer. He titled it 'Geology of Mankind', and it appeared in the prestigious science journal *Nature*. The idea of the Anthropocene immediately caught on. Since its introduction, it has encouraged an enormous amount of research and vigorous debate. It has triggered controversy and made demands on politicians. Above all, it has challenged the world to wake up to what we are doing to our planet.

In his groundbreaking 2002 paper, Crutzen mentioned the huge population growth over the past few decades and the demands that this is making on the planet's resources. He noted the rise in energy use and growth in production. We have been pumping more and more greenhouse gases into the atmosphere. He talked of the ozone-destroying chlorofluorocarbons (CFCs), the loss of forests on a scale never seen before, the exhaustion of ocean fish stocks, and the damage caused to land and sea by the industrial use of chemical fertilisers in agriculture. The relatively stable conditions of the Holocene seemed to be well and truly over.

* * *

However, before the idea of the Anthropocene can gain formal recognition, geologists have to be on board. They are the guardians of how geological time is divided, dated and named. And just in case there is any doubt about who is in charge, any suggestion of a new epoch named the Anthropocene first has to be considered and approved by the International Commission on Stratigraphy (ICS). This committee oversees and ratifies any changes to the Geological Time Scale. It needs evidence.

And evidence is now being sought and found. It is accumulating at a fascinating, if somewhat alarming, rate. Findings are pouring in from archaeologists, historians, geographers, chemists, earth system scientists and oceanographers, as well as geologists and palaeontologists. They all have to translate their findings into geologically recognisable phenomena so that future stratigraphers – the people who identify, name and date geological layers – will recognise that significant geological, mineralogical, biological and chemical changes have been taking place at this time, our time.

It has been estimated that well over 100,000 new, refined and manufactured metals, minerals and materials have been made by men and women, especially over the past few decades. These include unnaturally occurring concentrations of a whole range of metals, including iron and steel, titanium and vanadium, gold and especially aluminium, which was unknown before the nineteenth century.

Minerals that don't occur in nature begin to make their appearance in landfills. Indeed, it has been estimated that many landfill sites, chock-a-block with discarded mobile phones and other electronic gadgets, contain more valuable metals than traditional ore deposits.

Larry Greenemeier of Earthworks, an American environmental advocacy group, estimates that if just one million discarded mobile phones were recycled, 14.7 kilograms of gold could be recovered (reported in *Scientific American*). This is a level of concentration far greater than any gold mine can boast. Mining rubbish tips is seen as a distinct possibility, not just for silver and gold, but equally for precious lithium and some of the rare earths. Even more straightforward is to recycle discarded lithium batteries, both large and small. Experts in battery recycling remind us that as well as lithium, valuable cobalt, nickel and copper can also be recovered.

Tungsten carbide, artificial garnets for lasers, and boron nitride that is even harder than diamond, chemical solvents, pesticides, lead and mercury can be found contaminating sedimentary layers at the bottom of lakes and seas, swamp and peat bogs, and on the surface of ice sheets and desert sands. These polluted layers are likely to hang around for thousands of years.

In order to manufacture all of our metals and minerals, billions and billions of tons of earth and rock have to be scooped, scraped, mined and moved every year. Not even the ocean floors are immune from these ravages. Mining companies are taking an increasingly keen interest in dredging the beds of the sea for minerals, particularly cobalt, copper, manganese and nickel – metals that ironically play a key role in the 'green' technology revolution. Occupying two-thirds of the planet, the seabeds of the world, according to *National Geographic*, contain more valuable minerals and metallic nodules than all of the continents combined. As the mining companies drag their machines indiscriminately across the ocean floors, they destroy marine life, disrupt ecosystems and contaminate the waters.

'The transformation of the Earth's land surface by mineral extraction and construction,' writes the geologist Anthony Cooper and his colleagues, 'is on a scale hugely greater than natural erosive terrestrial geological processes... Humans are now the major global geological driving force and an important component of Earth System processes in landscape evolution.'

Shifting this amount of crustal material from one place to another outdoes nature's ability to weather and wear rocks away many times over. Huge mining corporations level mountains. We drain swamps and fill in lakes. Coastal margins are reclaimed and built over with cities, airports and docks. Artificial islands are created, built from sand dredged from the bed of the sea and then covered by millions of tons of concrete poured by the truckload.

Abandoned mines leave mile after mile of collapsing shafts that riddle the rocks deep beneath the surface. 'We are disembowelling the Earth,' says Gaia Vince, 'through mining, drilling, and other extractions.' Underground beds of ironstone, coal and gypsum come to an abrupt, unnatural end as ancient geological strata come up against the hollowed-out worlds left by tunnelling miners. Geological discontinuities usually map radical changes in depositionary conditions, or mark breaks in the passage of time during which rocks were weathered and worn away. But we create these sudden gaps, halts and unconformities every time we dig, tunnel, scrape and strip the Earth's crust. The geological anomalies and discontinuities of the Anthropocene, such as rock removal and the dumping of waste, might well pose a puzzle for geologists of the far future, if such a breed still exists.

One of the sharpest anthropogenic signals now being found worldwide is radioactive dust. Such dust was blasted high into the air every time a nuclear bomb was dropped or tested. This lethal leftover poison accumulates on lake floors, seabeds and ice sheets. Artificial isotopes of plutonium, caesium and carbon have all been detected where they would not naturally be found.

As we discussed in Chapter 9, plastic is turning out to be one of the world's biggest pollutants. The qualities that make it a technological wonder also make it an environmental nightmare. It is strong, durable and resists chemical breakdown. Plastic fragments, shredded plastic bags and microplastics from washing clothes made of synthetic fibres and from car tyre abrasion hang around for millennia. Plastics are dumped in rubbish tips, thrown into rivers, strewn along beaches and washed out to sea, where they eventually settle, along with muds, silts and sands, to form a distinct layer on the seabed and far out across the ocean floor. A study by the geologist Ian Kane and his colleagues found that in parts of the Mediterranean Sea, concentrations of microplastics are so extreme, not only are they polluting marine life, they are also forming a visible layer on the sea floor. Future geologists might wonder about these strange rock bands peppered with long-chain carbon polymers and plastics that have no known equivalent in nature.

Jennifer Brandon and colleagues at the Scripps Institute of Oceanography have studied the rise of plastic pollution in sediments laid down off the coast of California since 1834. However, it was in sediments laid down after 1945 that they found the biggest increases in such pollution, with rates rising exponentially. Most of the particles they found were plastic fibres from synthetic fabrics. Every time we wash our polyester shirts, our synthetic fabrics, our nylon socks, tiny microplastic fibres are released to make their way down waste pipes, through sewage works, into rivers and out to sea. Brandon says that our love of plastics is laying down a new fossil record, suggesting that future generations might name our time as the 'Plastic Age'.

New 'rock' types are also forming on land. Cities are built on layer upon layer of crumbling bricks, fragmented pots and broken glass. Hundreds of thousands of miles of tarmacked roads stretch their way between cities and across continents. What peculiar linear remains will they leave behind?

But it's concrete that has been identified as the signature 'rock' of the Anthropocene. Billions of tons of concrete are produced annually. It is poured to make dams, tunnels, bridges, skyscrapers, multi-storey car parks, roadways, pavements and foundations. And when its days are over and buildings are demolished, the crushed concrete doesn't disappear. It is left as rubble, landfill and urban waste. Just like natural sediments, it gets dispersed and buried. It accumulates in pockets and layers, where, in a million years' time or more, it could still be present as a strange calcareous, limey layer sandwiched between bands of sand or beds of shale or solidified seams of household rubbish and industrial waste.

Emissions of carbon dioxide from power stations, cement kilns and car engines are implicated in global heating. In turn, this leads to melting ice caps. For example, a study led by the geophysicist Ingo Sasgen found that in 2019 the Greenland ice sheet had lost a record amount of ice, equivalent to a million tonnes per minute across the year. In 2020, a large section of the Greenland sea ice shelf, known as N79, broke off and broke up as it slipped into the ocean. Ice loss has tripled since the turn of the twenty-first century, as documented by NASA. Melting ice leads to rises in sea levels. Higher seas flood low-lying coastal areas. They alter the geography and shift the pattern of sedimentary movements and deposition. All of these geomorphological changes will appear in the geological records of the future.

As well as producing carbon dioxide and other noxious gases, burning wood and fossil fuels releases smoke and soot. These tiny particles of fly ash and black carbon eventually settle over the land and fall into lakes and rain down on the sea. Trace layers of these particles have been found in recent sedimentary deposits worldwide. This global footprint of human activity is now regarded as another key signature of the Anthropocene.

* * *

Perhaps the most profound change of all is taking place in the 'biological signal'. If one of the key indicators of shifts in the planet's geological, climatic and biological character is to be seen in the fossil record, then what kind of record is being left in the Anthropocene? Explosions of new life forms and mass extinctions have often marked the boundary between one geological period and another. Could the pollutants and poisons being released by our mines and industries be killing life on a global scale? Is global warming sounding the death knell for untold numbers of plants and animals? Has chopping down more than half the world's trees since our arrival on the planet destroyed whole ecosystems? Does farming on an industrial scale, growing only a few crops, rearing only some animals, fertilising with synthetically manufactured nitrogen-based chemicals, signal the demise of many insects, birds and mammals? Are we on our way to *a sixth major mass extinction*? Are we seeing the first great extermination event, asks the environmental historian Justin McBrien, ecocide committed by just one species, Homo sapiens? Will palaeontologists of the future notice the sudden disappearance of tens of thousands of species from the fossil record as the Anthropocene gets under way? The signs are not looking good.

Insect numbers and diversity are falling dramatically. Microorganisms in the soil are being lost. Many bird populations are plummeting. Fish stocks are being depleted. Corals are dying. Rhinos are disappearing. Polar bears are threatened. The ecologists Gerardo Ceballos, Paul Ehrlich and Peter Raven found that more than five hundred species of land vertebrates are currently on the brink of extinction. Habitats suitable for wildlife are shrinking by the day.

The health of the biosphere appears to be in a very poor state. The May 2019 UN global assessment report warns that the planet's life-support systems are being destroyed by human activity. We destroy forests and grasslands to grow single crops and rear animals for meat, we degrade the land, we overfish, we pollute, and we burn fossil fuels and cause changes to our climate. *The State of the World's Plants and Fungi* 2020, published by the Royal Botanic Gardens, Kew, reports that 40 per cent of the world's plant species are currently under threat of extinction.

Across the plant and animal kingdoms, up to one million species, and counting, are at imminent risk of annihilation. And with their loss, we put ourselves in danger, too. Humanity's health and well-being rely on a rich biodiversity. Food production, crop pollination, clean water, land quality and human health all come under threat if we plough on regardless. Ceballos and his colleagues cite the coronavirus, COVID-19, pandemic as an example of what can happen when we ravage the natural world, trade in wildlife and destroy so many natural habitats.

There is another, parallel, story also being told in the biological record. Every year we rear and slaughter billions of chickens. Beef cattle graze the grasslands where extensive forests once grew. Future palaeontologists might well puzzle over the extraordinary number of fossilised chicken, cow and sheep bones found in man-made layers of rock rubbish the world over. If the Jurassic and Cretaceous periods have been dubbed the Age of the Dinosaur, might far-future generations examining the fossil record mockingly name our times the Age of the Chicken?

The geologist and palaeobiologist Jan Zalasiewicz has also calculated that there are now some 300 million domestic cats in the world – that is, roughly 100,000 house cats for each endangered tiger left in the wild. In the distant future, given the fossil record, it might seem that cats, along with dogs, had been, and might still be, a remarkably successful worldwide species.

Things are no different in the plant word. Single cereal crops of wheat, rice and maize have replaced the diverse flora that once painted plains, prairies and river valleys in bright colours. Rainforests and temperate woodlands and all the wildlife they support have dwindled as palm oil and grain crops take their place. When examined in the far future, the types and numbers of pollens and tree spores settling on the bottom of lakes and lagoons will seem to have fallen overnight.

* * *

There are many Earth scientists who believe that the evidence is now strong enough to suggest that the Anthropocene is, indeed, a real phenomenon. 'The human reign,' writes Jan Zalasiewicz, 'has been geologically brief – but then so was the meteorite impact that likely ended the world of the Cretaceous and ushered in the Cenozoic Era. In this case, we collectively are (in effect) the meteorite.'

The next question to tax scientists is, if the Anthropocene is real, when did it begin? Many suggestions have been made. In their book *The Human Planet: How We Created the Anthropocene*, Simon Lewis and Mark Maslin offer a detailed analysis of how human beings have affected and continue to affect the planet, chemically, atmospherically, biologically and geologically.

The transition from hunter-gatherers to settled farmers saw people cut down trees, grow cereal crops and domesticate a small number of species.

The second transition began in the sixteenth century. It involved the people of western Europe beginning to colonise and trade with the rest of the world. This also meant that some species, both plants and animals, that for millions of years had lived and evolved separately on different continents, were being introduced, by both design and accident, to other continents. This not only upset and altered local ecosystems, it represented a sudden change in the flora and fauna that might at some future date be recorded in the fossil record. There was both a global homogenisation and simultaneous reduction of plant and animal species. For the first time there was a large-scale intercontinental swapping of people, animals, plants and diseases. Genetic diversity and distinctiveness were being lost. A reordering of life on Earth was taking place, and its cause was human exploration, expansion and exploitation.

Lewis and Maslin therefore suggest that the year 1610 might be a good time to mark the beginning of the Anthropocene. They observe that it was the last globally cool moment before the Earth began its long-term, humanity-driven phase of warming.

They also argue that a short-term period of cooling had been caused by the arrival of Europeans in the Americas. In acts of genocide, the invaders killed tens of thousands of indigenous American peoples. The Europeans also exposed native populations to new diseases to which they had no immunity, leading to the deaths of millions more. Many of the American people were farmers, and when their numbers fell and their farming stopped, trees returned to the pastures, capturing carbon dioxide from the atmosphere. This resulted in a small, but measurable, dip of 7 ppm of the greenhouse gases, leading to a slight global cooling. However, this didn't last long as the colonisers themselves were soon clearing land to create farms from the North American east coast to the west.

Lewis and Maslin believe that these changes in carbon dioxide levels and global temperatures, and the reduced biodiversity, will leave a clear mark in the geological record. Such moments in the geological record are known as 'golden spikes' and so Lewis and Martin have called the year 1610 the 'Orbis Spike' from the Latin word for world, *orbis*.

The third transition began in the eighteenth century, as people moved from farming the land to working in factories. Labour became organised. Cities grew. Machines, at first powered by water, were then driven by steam and to make steam you had to burn coal. As we learned, by the late 1770s, the Scottish inventor James Watt had improved Newcomen's original steam engine. Many see Watt's engine as the real beginning of the Industrial Revolution. Coal mining and coal burning grew and grew. But this was also the beginning of rising levels of carbon dioxide. Man-made global warming and climate change were underway. The late eighteenth century is therefore another strong candidate for marking the beginning of the Anthropocene.

However, many believe that the middle of the twentieth century is the best place to start. This is when mining and manufacturing, invention and innovation, production and consumption, people and populations began to grow at faster and faster rates. The impact of all of this accelerated activity is profound. It is leaving signals and scars in the landscape, at the bottom of lakes, over seabeds and in the air.

Since the 1950s, the human population has tripled, carbon dioxide emissions have soared, average global temperatures have risen, sea levels are higher, concrete production has grown hugely, atomic bombs have been exploded, mining for metals has expanded, and plastics are everywhere. All of these factors have left their mark in and on the planet's surface. There is a new geological imprint and it is man-made. Millions of years from now, says Gaia Vince, there will be a stripe in the rocks and that stripe will have been made by us. The stripe is saying that after this time, the Earth moved into a new state and it was caused by human beings.

Scientists who are members of the Anthropocene Working Group have convinced themselves that the changes do represent a geological reality. They continue to collect the evidence. However, whether the time lords of the International Commission on Stratigraphy will be convinced is yet to be seen. They gaze over billions of years of Earth's history. Humankind's presence on the Earth is a mere blink in geological time. Nevertheless, all are agreed that human beings are having a radical impact on the planet – physically, climatically and biologically – regardless of whether or not it will lead to a signal in the geological record sufficient to be recognised in some distant future.

* * *

In these pages we have looked at how people have learned to turn rocks into stuff. The things that we make vary from the massive and grand to the tiny and discreet. Concrete dams and towering skyscrapers of steel and glass create entirely new landscapes. At the other end of the manufacturing spectrum, minute amounts of gold, platinum and many of the rare earth metals have made possible the IT revolution as they carry out their magic inside our phones, computers and television sets.

I began this story of stuff remembering my amazement as a teenager as I realised that as a species we had learned how to transform the earth beneath our feet into the world of things around us. Nature was certainly bountiful, but it took the genius of men and women to work out how to get iron, copper, silica and aluminium out of the rocks, and then, having achieved that, to go one stage further and generate electricity from the iron and copper, make plates of glass from grains of sand, and press sheets of aluminium to shape the wings of an aeroplane. These technological miracles are testimony to the cleverness of our species.

But maybe sometimes we are too clever for our own good. It is dawning on us that a price is being paid for our ingenuity. Although some of the Earth's resources seem inexhaustible, others are finite and limited. It takes millions of years for fossil fuels to form, but we are burning them at a rate millions of times faster than they can ever be replaced.

Nature's willingness to concentrate some metals and their minerals in economically worthwhile amounts is not boundless. The precious metals are precious because they are rare and hard-won. Lithium, silver and the rare earth elements that have helped power the Age of Information Technology occur in only a few places where they can be realistically mined. Although it may be some time off, there will come a day when many of the metals we mine won't be there any more.

It's true that we have shown a remarkable ability to engineer our way out of the concerns and worries of the doomsayers. We have invented new technologies when old ones seemed on the point of failure. We have created new materials when old ones were no longer available. We are learning how to harness the power of wind, water and sunlight to fuel a 'greener' world. We can use electricity generated by renewable technologies to electrolyse water to produce hydrogen, which can be used to store energy and power cars, trains and even aeroplanes. We are finding ways to capture carbon from the atmosphere and return it to the rocks of the upper crust. And there's no doubt we shall continue to invent, create and develop our way around setbacks and shortfalls as we have been doing for thousands of years.

But the current scale of our assault on the planet and our rapacious appetite for rocks and minerals is like nothing else before. The past couple of hundred years have seen our hunger for coal, oil, iron ores, copper ores, bauxite and the thousands of other rock types and minerals we mine grow enormously. What has taken nature hundreds of millions of years to contrive and concentrate, we are exploiting, possibly to the point of exhaustion, in just a few years, barely a geological moment. Our insatiable appetite for fossil fuels and metallic ores is a geological experiment that can be carried out only once in our species' lifetime.

* * *

It was during the 1960s that James Lovelock came up with the idea of the Earth as a self-regulating, living system. Lovelock's neighbour and friend, the novelist William Golding, suggested the scientist name his hypothesis Gaia after the Greek goddess of the Earth. Lovelock proposed that living organisms interact with their physical environment to maintain planetary conditions suitable for life and its perpetuation.

Feedback loops between life, land, air and water ensure that oxygen, temperature and light radiation levels stay within limits suitable for plants and animals. When one of nature's elements or cycles starts to get out of kilter, disturbing the planet's homeostatic equilibrium, other natural elements and cycles work to bring it back. In this way, conditions for life on Earth are continually optimised. It was a revolutionary idea that saw links between biology, geology and the Earth's chemistry.

Gaia theory has been criticised by some scientists as being too mystical, as lacking a strong evidence base, as seeming more metaphor than scientific theory. Lovelock himself has courted controversy over his changing views on the causes, reality and future of climate change. Nevertheless, the originality of his thinking helped bring about a willingness to look at the Earth as a whole system, whether or not it was one that actually had a built-in mechanism to ensure that life interacted with the planet in such a way as to maintain its own survival. What has been established is that changes in one part of the Earth's system usually trigger changes in other parts of that system, for good or ill. There is no underlying mechanism that necessarily favours life, although living organisms can and do react in ways that attempt to mitigate many otherwise biological adverse changes. But success is not guaranteed.

What environmental and Earth scientists are revealing are some of the consequences, intended and unintended, of what happens when we alter, shift and plunder the planet to meet our own needs. For example, there is no getting away from the fact that no matter how wonderful plastics are, unless we behave more responsibly, they will poison the seas, kill marine life and choke birds.

Here is another set of consequences. As we've seen, burning coal, oil and gas produces carbon dioxide. Making cement also creates carbon dioxide. The gas increases the acidity of the sea, which upsets sea life. Carbonate minerals dissolve in weak carbonic acid. Acidic seas mean that animals with skeletal structures and shells made of calcium carbonate find it difficult to grow. Corals die. Fewer fish larvae survive. Marine ecosystems crash.

Carbon dioxide is also a greenhouse gas. There is now no denying that average global temperatures are warming. Climate change is a reality. A hotter world has consequences, for geography, geology, biology, agriculture, economies and politics. The world warms and habitats change or disappear and along with them the lives they sustain. Ice melts. Sea levels rise. Coastal lands, where many of the world's population live, are lost to the sea. Oceans get warmer and cold-water species and those that feed off them die.

A warmer, more energetic atmosphere leads to bigger storms and, in some parts of the world, higher rainfall. Heavy rains increase the risk of floods and soil erosion.

In other parts of the world, summer temperatures become unbearably hot. Lands that were once fertile turn to dust and blow away, and deserts advance. Failed crops lead to famines, and famines put pressure on desperate people to move and migrate. Tens of millions of climate refugees are set to walk the planet. Population movements lead to political problems, overcrowding, cultural clashes and conflict.

And so the knock-on effects go on. That's what happens when we burn millions of years of fossilised sunlight in a couple of human lifetimes. It all adds up to what the novelist Amitav Ghosh calls 'the great derangement', a thought echoed by the 2014 report of the American Association for the Advancement of Science (AASS) Climate Science Panel, *What We Know: The Reality, Risks, and Responses to Climate Change*, chaired by Mario Molina:

Most projections of climate change presume that future changes – greenhouse gas emissions, temperature increases and effects such as sea level rise – will happen incrementally... However, the geological record for the climate reflects instances where a relatively small change in one element of climate led to abrupt changes in the system as a whole. In other words, pushing global temperatures past certain thresholds could trigger abrupt, unpredictable and potentially irreversible changes that have massively disruptive and large-scale impacts.

Those are the kind of stories Earth system scientists tell when they are feeling pessimistic. But they also do optimism. The more positive side of their work helps us gain a better understanding of what is likely to happen when we do things to the planet, when we mine, manufacture and manipulate the Earth and its resources. It does not necessarily mean that we should stop doing these things, but having a better understanding can help mitigate their worse effects, give us alternative ideas or warn us of the environmental dangers ahead.

If you must burn coal (but we'd rather you didn't), then think about capturing the carbon dioxide and preventing it from entering the atmosphere. Car engines and train engines shouldn't be driven by oil and its products. Electric cars, battery technology and hydrogen-powered engines are the way forward, powered ultimately by green energy from the sun, wind, tidal seas and tumbling waters. Allowing the regrowth and rewilding of mangrove swamps, salt marshes, tropical forests, temperate forests, peat bogs and mountain woodlands is a natural way of capturing atmospheric carbon dioxide.

Plastics have so many uses. It's impossible to imagine our lives without them. But make them recyclable or out of plant cellulose. Make them biodegradable.

In principle, all metals are endlessly recyclable. The minerals from which we smelt them are not inexhaustible. The Earth's crust has only so much copper, lithium, platinum and silver. So don't throw away your mobile phone and its tiny cargo of gold, silver, copper and all of those rare earth elements. Collect them, scavenge them, gut them, remelt them, reuse them. One metric tonne of silver ore will yield at most just 3 grams of the metal. One metric tonne of dead mobile phones can contain as much as 3 kilograms of silver. Gaia Vince notes that most 'electronic waste' contains more precious metals than the ores from which they were originally mined. And even the more plentiful metals are costly to manufacture, using huge amounts of energy in the process. Don't dump aluminium cans into waste tips – recycle them. There's value in scrap iron. Throw it back into the blast furnaces.

Whenever possible, runs the environmentalist's mantra: *renounce, reduce, reuse and recycle.*

In *The Story of Stuff,* Annie Leonard reminds us of the sheer wastefulness of the way we organise our materials economy. The system is linear. We go from extraction to production to distribution to consumption to disposal. Every stage has implications for people and the environment.

Linear systems will not work forever on a finite planet. Modern economic systems of deregulated capitalism demand that we make consumption our way of life. Capitalism requires that we buy, discard and replace at an ever-increasing rate. It demands perpetual growth. Its appetite for the planet's resources knows no limits. But the world's resources, geological and biological, are ultimately limited. We live, says the French thinker Bruno Latour, in a narrow zone on the Earth's surface. He calls this the 'critical zone', with a band of air above and surface crust below, a zone that supports all life. 'In the critical zone,' he says, 'we must maintain what we have because it is finite, it's local, it's at risk and it's the object of conflict.'

All of this is discussed and analysed by the economist Partha Dasgupta in his major review, *The Economics of*

Biodiversity. He argues that the air, water, rocks, soil and all living things have to be seen as 'natural capital' and that humanity is running down this natural capital at an alarmingly fast rate. He writes that, 'We are part of Nature, not separate from it.' Fundamentally, it is nature that provides us with everything we need to live, from the air we breathe, the water we drink and the food we eat to the raw materials we use to make a world of so much manufactured stuff.

Nature is an asset, but one that we cannot take for granted. If we continue to plunder and exploit nature without regard, it will be future generations who pay the price as they struggle to survive in a hotter, more polluted, less resource-rich, less biodiverse planet. Our economic systems need to be truly sustainable. We need to decarbonise our energy systems. Our long-term well-being, indeed our survival, depends on rebalancing our relationship with nature. In the report's headline messages, Dasgupta concludes:

Humanity faces an urgent choice. Continuing down our current path – where our demands on Nature far exceed its capacity to supply – presents extreme risks and uncertainty for our economies. Sustainable economic growth and development requires us to take a different path, where our engagements with Nature are not only sustainable, but also enhance our collective wealth and well-being and that of our descendants.

The Canadian-American writer and activist Astra Taylor also warns that conventional business models worship the financial bottom line and short-term thinking. She says, 'Their timescales privilege the present.' They are impatient with nature's pace. They close their eyes to the interconnectedness of the planet's biological, chemical and physical systems.

Economies based on capitalism constantly need to grow and expand. Capitalism devours nature and gobbles the Earth's resources. It excretes waste and values obsolescence; it pollutes and degrades. It is a beast we ride but cannot control. Once mounted, we discover it has a life and logic of its own. 'Capitalism,' says Taylor, 'lacks the attention span required for survival,' while the Austrian writer and journalist, Karl Kraus reflects wryly that 'Progress celebrates Pyrrhic victories over nature.'

In his 1970 review of Max Nicholson's book *The Environmental Revolution*, the poet Ted Hughes praised an approach that took 'nature's point of view'. There is a 'wholeness' to nature and the universe that we ignore at our peril. In his review, Hughes despaired of modern capitalism and the way it exploits the Earth. He wrote:

> While the mice in the field are listening to the Universe, and moving in the body of nature, where every living cell is sacred to every other, and all are interdependent, the Developer is peering at the field through a visor, and behind him stands the whole army of madmen's ideas, and shareholders, impatient to cash in the world.

Whether based on capitalism, state communism or centrally controlled planning, economic systems that rely on growth that doesn't know how to stop will inevitably lead to environmental calamity. The author and social activist Naomi Klein puts it even more starkly: 'Our economic systems and our planetary systems are now at war.' As we kill the planet, the planet's complex feedback loops will end up killing us as we degrade the land, increase greenhouses gases and destroy biodiversity. This is how Lewis and Maslin in their book, *The Human Planet*, characterise the perverse logic:

> *The capitalist mode is directed to increasing profits, which ultimately means increasing the productivity of people. Energy, while critical, can be seen as a means to increase productivity. This central tenet is seen today: workers can only be paid more if their productivity increases. The result is that ever more is produced, then ever more must be consumed, which is at the core of the environmental changes seen in the post-war Great Acceleration period.*

Our fixation on growth, argues the economist Kate Raworth, has to stop. For too long, economic models based on relentless growth have been used to justify extreme inequalities of income and wealth, extracting minerals without regard for the environment, and increasing consumerism to support increased productivity.

The fact is that the more wealth you have, the greater your impact on the environment. You consume more, burn more and discard more. In Chad, each person generates about 0.06 metric tonnes of carbon a year. In India the figure rises to 1.91 tonnes per person per year, while in the UK the amount rises to 5.48 tons per year. But in the USA, each person, each year, produces 16.1 tons of carbon, a carbon footprint 268 times greater than that created by each citizen of Chad. And yet the richer you are, even though you are disproportionately responsible for the crises affecting the environment, the less you are likely to suffer as the world overheats, endures droughts, suffers floods and runs out of resources. It is the poor who face disaster as their lands dry up and disappear, their resources shrink and prices rise.

At the moment, Africa contributes less than 4 per cent to global carbon dioxide emissions, but it is the continent most likely to be hardest hit by the disruptive effects of global heating. The prospect of a 'climate apartheid', writes the environmental journalist Fiona Harvey, 'in which the rich insulate themselves from the impacts of global heating while the poor are abandoned to their fate, is all too real'.

The extreme weather patterns triggered by global heating will lead to increasingly unstable social conditions. Economic and political systems will be put under mounting strain as millions of desperate people attempt to flee famine, fire and flood.

In response, Annie Leonard argues for more durable products that last longer. Instead of being thrown away when they don't work, devices should be made repairable. We should not only recycle more but also waste less. She argues for a more circular materials economy, one that uses less energy and that reuses and recycles more. Mimicking nature, we need closed-loop production systems.

Biomimicry seeks design solutions inspired by nature. Over hundreds of millions of years, nature has engineered biological systems that are self-sustaining. We can learn from nature. Sunlight, wind and water power can meet most of our energy needs. Biological systems use water-based chemical reactions to run themselves rather than toxic mixes of dangerous chemicals. In nature, everything gets recycled.

There is a growing recognition that if we are to avoid environmental ruin, we should accept the notion of 'private sufficiency, public luxury'. The planet cannot afford the rich who consume, exploit and expend at rates hundreds of times greater than the world's poor. 'We should adopt a new conception of justice,' says the writer and environmental activist George Monbiot, 'based on the principle: every generation, everywhere, shall have an equal right to the enjoyment of natural wealth.' Sustainability is the watchword. Life on Earth depends on moderation. We must put a limit on wealth.

In his final 2020 Reith Lecture in which he talks about the climate crisis, the economist Mark Carney warns governments, business and global capital that, 'For an issue that can only be solved in the present, we have to value the future.' He goes on to say that as a species we have traded off the planet and its resources without giving any thought to future generations. We have been greedily living for today and leaving it to others to pay for it tomorrow. Carney argues that by revaluing what fundamentally and existentially matters to people and the planet, governments need to give companies new incentives that reward becoming part of the solution, and making it very costly if they persist in remaining part of the climate crisis problem. When governments unambiguously support 'green' technologies and demand resource sustainability, the more likely it will be that companies will invest in activities that are environmentally efficient, innovative and enterprising.

Throughout history, Nemesis, the goddess of retribution, has snapped at the heels of human hubris. Our wish to survive and thrive requires us to respect the Earth and its gifts. It demands that we understand we are just one thread in the complex web of life and that whatever we do, there are consequences, not just for other species, but for us too. The Earth is a system of densely woven, interrelated feedback loops. When we act in a way that is detrimental to the planet and the living things that it supports, somewhere along the endless line of action and reaction, we end up reaping what we sow. And what we reap may not always be good for us.

Yes, we are technologically clever, but all too often we can be environmentally stupid. We shall continue to mine and extract, manufacture and make, use and consume, but we have to listen to what environmental scientists have to say and heed the warnings of the ecologically wise. We have to consume less. The footprints we leave on the planet have to be smaller. Our economies have to be circular and not linear. Whatever we do, the Earth will still be here. But whether we will be around to see it will depend on us.

References

American Association for the Advancement of Science (2014), *What We Know: The Reality, Risks, and Responses to Climate Change*, Washington, DC: AAAS.

Arndt, N., Kesler, S. and Ganino, C. (2010), *Metals and Society: An Introduction to Economic Geology*, New York: Springer.

Barthes, R. (1972), *Mythologies*, London: Jonathan Cape.

Brandon, Jennifer A., Jones, W. and Ohman, M. D. (2019), Multidecadal increase in plastic particles in coastal ocean sediments, *Science Advances*, Vol. 5 No. 9, pp. 1–6.

British Geological Survey (2009), *Silica Sand: Mineral Planning Fact Sheet*, London: NERC.

British Geological Survey (2011), *Rare Earth Elements*, London, NERC.

British Geological Survey (2016), *Lithium*, London, NERC.

Bronowski, Jacob (1973), *The Ascent of Man*, London: BBC Publications.

Bryson, Bill (2019), *The Body: A Guide for Occupants*, London: Doubleday.

Carney, Mark (2020), *The Reith Lectures, Episode 4: From Climate Crisis to Real Prosperity*, BBC Radio 4, 23 December.

Cebellos, G., Ehrlich, P. R. and Raven, P. (2020), *Proceedings of the National Academy of Sciences*, 1 June: https://doi.org/10.1073/pnas.1922686117 (accessed 21/1/21).

Cheng, L. et al. (2020), Record-setting ocean warmth

continued in 2019, *Advances in Atmospheric Sciences*, February, Vol. 37, No. 2, pp. 137–142.

Cobb, Harold M. (2010), *The History of Stainless Steel* (Chapter 5: 'The Life of Harry Brearley'), Ohio: ASM International, pp. 33–57.

Cooper, Anthony H. et al. (2018), Humans are the most significant global geomorphological driving force of the 21st century, *Anthropocene Review*, Vol. 5, No. 3, pp. 222–229.

Crutzen, P. J. (2002), Geology of mankind, *Nature*, Vol. 415, p. 23.

Crutzen, P. J. and Stoermer, E. F. (2000), The 'Anthropocene', *Global Change Newsletter*, 41, 17.

Dartnell, Lewis (2019), *Origins: How the Earth Made Us*, London: The Bodley Head.

Dasgupta, P. (2021), *The Economics of Biodiversity: The Dasgupta Review*, London: HM Treasury.

Economist, The (2012), 'The lore of ore', 13 October: https://www.economist.com/finance-and-economics/2012/10/13/the-lore-of-ore (accessed 21/1/21).

Elhacham, E., Ben-Uri, L., Grozovski, J. et al. (2020), Global human-made mass exceeds all living biomass, *Nature*, Vol. 588, pp. 442–444: https://doi.org/10.1038/s41586-020-3010-5 (accessed 21/1/21).

Evans, Anthony M. (1997), *An Introduction to Economic Geology and its Environmental Impact*, Oxford: Blackwell.

Fortey, Richard (2004), *The Earth: An Intimate History*, London: HarperCollins.

Fortey, Richard (2010), *The Hidden Landscape: A Journey into the Geological Past*, London: The Bodley Press.

Forty, Adrian (2012), *Concrete and Culture: A Material History*, London: Reaktion Books.

Freinkel, Susan (2011), *Plastic: A Toxic Love Story*, Boston: Houghton Mifflin Harcourt.

Gardiner, B. (2019), *Choked: The Age of Air Pollution and the Fight for a Cleaner Future*, London: Granta Books.

Ghosh, Amitav (2016), *The Great Derangement: Climate Change and the Unthinkable*, Chicago: University of Chicago Press.

Gill, A. A. (2015), *Pour Me: A Life*, London: Weidenfeld and Nicolson.

Gray, John (2002), *Straw Dogs: Thoughts on Humans and Other Animals*, London: Granta Books.

Greenemeier, Larry (2007), Trashed tech: where do old cell phones, TVs and PCs go to die?, *Scientific American*, 29 November.

Greenpeace (2019), *Ghost Gear: The Abandoned Fishing Nets Haunting Our Oceans*, Hamburg: Greenpeace.

Harvey, Fiona (2019), 'Adaptation alone won't save us from disaster', *The Guardian, Journal*, 12 September, pp. 1–2.

Heede, Richard (2019), Carbon Majors: Updating activity data, adding entities, and calculating emissions: A training manual, September, Snowmass, Colorado; Climate Accountability Institute, p. 56.

Higgins, Polly (2010), *Eradicating Ecocide: Laws and Governance to Stop the Destruction of the Planet*, London: Shepheard-Walwyn.

Higgins, Polly (2011), *Earth Is Our Business: Changing the Rules of the Game*, London: Shepheard-Walwyn.

Hughes, Ted (1970), 'An idea whose time has come', review of Max Nicholson's book *The Environmental Revolution*, in *Your Environment*, Vol. 1, No. 3.

Kane, Ian A. et al. (2020), Seafloor microplastic hotspots controlled by deep-sea circulation, *Science*, 5 June, Vol. 368, No. 6,495, pp. 1140–1145.

Kew (2020), *State of the World's Plants and Fungi 2020*, London: Royal Botanic Gardens, Kew.

Klein, Naomi (2014), *This Changes Everything: Capitalism vs. the Climate*, New York: Simon and Schuster.

Kraus, Karl and Zohn, Harry (1990), *Half Truths and One-and-a-Half Truths: Karl Kraus, Selected Aphorisms*, Chicago: The University of Chicago Press.

Latour, Bruno (2020), Interview, *The Observer, Review*, 7 June, p. 20.

Lau, Winnie W.Y. et al. (2020), Evaluating scenarios toward zero plastic pollution, *Science*, July, Vol. 369, No. 6,510.

Leeder, Mike and Lawlor, Joy (2015), *GeoBritannica: Geological Landscapes and the British Peoples*, Edinburgh: Dunedin.

Leonard, Annie (2010), *The Story of Stuff: The Impact of Over-consumption on the Planet, Our Communities, and Our Health – and How We Can Make It Better*, New York: Free Press.

Lewis, Simon L. and Maslin, Mark A. (2018), *The Human Planet: How We Created the Anthropocene*, London: Pelican Books.

Lovelock, James (1979), *Gaia: A New Look at Life on Earth*, Oxford: Oxford University Press.

Malina, M. et al. (2014), What We Know: The Reality, Risks and Response to Climate Change, *AAAS Climate Change*

Panel, American Association for the Advancement of Science, pp. 15–16.

Malm, Andreas (2015), *Fossil Capital: The Rise of Steam-Power and the Roots of Global Warming*, London: Verso Books.

Maloney, F. J. Terence (1967), *Glass in the Modern World*, London: Aldus Books.

McBrien, Justin (2019), This is not the Sixth Extinction. It's the First Extermination Event, *Truthout*, 14 September.

Miodownik, Mark (2014), *Stuff Matters: The Strange Stories of the Marvellous Materials that Shape Our Man-made World*, London: Penguin.

Monbiot, George (2019), 'Dare to declare the system dead – before it takes us down with it', *The Guardian, Journal*, 25 April, pp. 1–2.

Monbiot, George (2019), 'For the sake of life on Earth, we must put a limit on wealth', *The Guardian, Journal*, 19 September, pp. 1–2.

Nash, David et al. (2020), Origins of the sarsen megaliths at Stonehenge, *Science Advances*, 29 July, Vol. 6, No. 31.

National Oceanic and Atmospheric Administration (NOAA), Earth System Research Laboratory: Global Monitoring Division, Boulder, CO: US Department of Commerce.

Nenquimo, Nemonte (2020), 'Western living is killing life on Earth', *The Guardian, Journal*, 12 October, p. 4.

Nicholson, Max (1970), *The Environmental Revolution*, London: Hodder and Stoughton.

Ocean Conservancy (2018), *Building a Clean Swell*, Report, Washington D.C.

Pabortsava, K., and Lampitt, R. S. (2020), High concentrations of plastic hidden beneath the surface of the Atlantic Ocean, *Nature Communications*, 18 August, Vol. 11, Article 4,073.

Pearson, Mike Parker et al. (2019), Megalith quarries for Stonehenge's bluestones, *Antiquity*, February, Vol. 93, No. 367, pp. 45–62.

Pipkin, B. W., Trent, D. D. and Hazlett, R. (2005), *Geology and the Environment*, 4th edition, Belmont, CA: Brooks/Cole.

Pohl, Walter L. (2011), *Economic Geology: Principles and Practice Metals, Minerals, Coal and Hydrocarbons – Introduction to Formation and Sustainable Exploitation of Mineral Deposits*, Oxford: Wiley-Blackwell.

Rado, Paul (1969), *An Introduction to the Technology of Pottery*, London: Pergamon Press.

Raworth, Kate (2017), *Doughnut Economics: 7 Ways to Think Like a 21st Century Economist*, White River Junction, Vermont: Chelsea Green Publishing.

Rickard, T. A. (1939), *The Primitive Smelting of Iron*, American *Journal of Archaeology*, January–March, Vol. 43, No. 1, pp. 85–101.

Roszak, Theodore (1995), *The Making of a Counter Culture*, Berkeley: University of California Press.

Ruskin, John (1859), *The Two Paths*, London: George Allen.

Sasgen, I. et al (2020), Return to rapid ice loss in Greenland and record loss in 2019 detected by the GRACE-FO satellites, *Communications, Earth and Environment*, 20 August, Vol. 1, Article No. 8.

Shah, Sonia (2020), *The Next Great Migration: The Story of Movement on a Changing Planet*, London: Bloomsbury.

Stewart, M. et al (2020), Human footprints provide snapshot of last interglacial ecology in the Arabian interior, *Science Advances*, 18 September, Vol. 6, No. 38.

Taylor, Astra (2019), 'Burning legacy', *The Guardian, Journal*, 1 October, pp. 9–11.

Taylor, Astra (2019), *Democracy May Not Exist but We'll Miss It When It's Gone*, Brooklyn, NY: Verso.

Tennyson, Lord Alfred (1850), *In Memoriam,* London: Edward Moxen.

Vince, Gaia (2014), *Adventures in the Anthropocene: A Journey to the Heart of the Planet*, London: Chatto and Windus.

Wallace-Wells, David (2019), *The Uninhabitable Earth: A Story of the Future*, London: Allen Lane.

Watts, Jonathan (2019), 'Wicked problems are heading our way', *The Guardian*, 31 December, pp. 4–5.

Woodcock, Nigel and Strachan, Rob (2012), *Geological History of Britain and Ireland*, Oxford: Wiley-Blackwell.

Woodforde, John (1976), *Bricks to Build a House*, London: Routledge and Kegan Paul.

World Coal Association (2018), *Uses of Coal*, London: WCA.

Zalasiewicz, J. et al. (eds) (2019), *The Anthropocene as a Geological Time Unit: A Guide to the Scientific Evidence and Current Debate*, Cambridge: Cambridge University Press.

Zalasiewicz, J., Waters, C., Summerhayes, C. and Williams, M. (2018), The Anthropocene, *Geology Today*, Vol. 34, No. 5, pp. 177–181.

Acknowledgements

A big thank you to my publisher, Sara Hunt, for support-ing my second book for her. Fact-filled books must be a nightmare to edit, so I am extremely grateful to Ali Moore for going through the manuscript with such care and eagle-eyed attention, and to Yennah Smart for the proofread. I am delighted with the cover design, which is thanks to Daniel Benneworth-Gray. Geologists have an old saying that nearly became my working title for the book: 'If you can't eat it, it was probably found by a geologist.' Thanks to my friend Professor Godfrey Fitton for reminding me of this insight. Of course, any remaining errors or falsehoods in the book are entirely mine.

Much of this book was written during the COVID-19 pandemic. There are three other people I would especially like to thank: my daughter, Rebecca, and granddaughters, Elsa and Lucy. It was in early February 2020 when we found ourselves walking over Loughrigg Fell in the Lake District. The wind was blowing a gale and a horizontal rain was sting-ing our faces. They asked me, 'What are you writing at the moment, Grandad?' Against the noise of the wind I splut-tered something about rocks ending up as stuff – remarkably. But I said that our ingenuity also had downsides, including pollution, global warming and the loss of wildlife. The words 'extraction' and 'extinction' cropped up in the discussion. We liked the alliteration and so a possible title began to offer itself. Of course, what Elsa and Lucy would really like me to write are children's stories. Ever since they were young,

I've been sending them the occasional poem and a series of stories entitled *Tales of the Fearless Rob Jackson*. I've tried to explain that getting children's stories published is very competitive, but still they chivvy and challenge. Anyway, all you publishers of children's fiction out there, you know where to find me.

* * *

DAVID HOWE OBE is Emeritus Professor at the Centre for Research on Children and Families at the University of East Anglia. After starting out in Earth sciences, his academic career has focused in social sciences. He has written books and papers on psychology, relationships and social work as well as geology and landscapes. Now formally retired, he continues to lecture nationally and internationally. His passions include walking, popular science, and writing.

Index